4th International Workshop on Accelerator Radiation Induced Activation (ARIA 2017)

Journal of Physics: Conference Series Volume 1046

Lund, Sweden
22 – 24 May 2017

ISBN: 978-1-5108-6495-5
ISSN: 1742-6588

Printed from e-media with permission by:

Curran Associates, Inc.
57 Morehouse Lane
Red Hook, NY 12571

Some format issues inherent in the e-media version may also appear in this print version.

Printed by Curran Associates, Inc. (2018)

For permission requests, please contact the Institute of Physics at the address below.

Institute of Physics
Dirac House, Temple Back
Bristol BS1 6BE UK

Phone: 44 1 17 929 7481
Fax: 44 1 17 920 0979

techtracking@iop.org

Additional copies of this publication are available from:

Curran Associates, Inc.
57 Morehouse Lane
Red Hook, NY 12571 USA
Phone: 845-758-0400
Fax: 845-758-2633
Email: curran@proceedings.com
Web: www.proceedings.com

Table of contents

Volume 1046

4th International Workshop on Accelerator Radiation Induced Activation (ARIA 2017)

22–24 May 2017, Lund, Sweden

Accepted papers received: 31 May 2018
Published online: 12 June 2018

Preface

Papers

Reaction models. Programs and Data libraries

Cross section measurements and intercomparison

Benchmarking and intercomparison

Radiological characterization: Methods

Design studies

Environment

IOP Conf. Series: Journal of Physics: Conf. Series **1046** (2018) 011001 doi:10.1088/1742-6596/1046/1/011001

The fourth **International Workshop on Accelerator Radiation Induced Activation** (ARIA'17)

May 22-24, 2017 in Lund, Sweden.

The workshop follows ARIA'08 held at PSI, Switzerland in 2008, ARIA'11 held at Kibbutz Ma'ale HaChamisha, Israel in 2011 and ARIA'15 held at Knoxville, USA in 2015.

The purpose of these topical ARIA meetings is to provide an international forum for discussing the various problems of radionuclide transmutation in various applications at accelerator facilities.

The workshop provides an opportunity for nuclear physicists, accelerator physicists, nuclear engineers, and other experts in the international community to meet and discuss directly their research activities. These interactions can help establish good working relationships and collaborations to solve common problems across multiple disciplines related to neutron- and accelerator-induced activation.

The technical program will consist of a consecutive session of oral and poster presentations.

Scientific Programme

The workshop concerns those interested in neutron- and accelerator-induced activation, radioactive waste generation, transmutation and burnup, medical isotope production and radionuclide production in accelerator-driven physics facilities.

The main objectives of the ARIA meetings are to:

- Promote the exchange of information among experts in the field of neutron- and accelerator-induced activation, radioactive waste generation, transmutation and burnup, medical isotope production and radionuclide production in accelerator-driven physics facilities, and related topics.
- Identify areas where international co-operation can be fruitful in order to achieve progress in specific priority areas.

Topics of discussion include:

- Methods of predicting the production of radionuclides
- Methods of simulating activation of radiation fields
- Reaction cross section measurements
- Benchmarking of reaction models
- Developments in activation cross section libraries
- Radiation protection and health physics
- Environmental assessments
- Accelerator operations and maintenance issues
- Waste management
- Radioisotopes applications

IOP Conf. Series: Journal of Physics: Conf. Series **1046** (2018) 011002 doi:10.1088/1742-6596/1046/1/011002

ARIA17

International Organizing Committee:

- Daniela ENE, ESS/Sweden
- J.-Ch. SUBLET, UKAEA/Great Britain
- Franz GALLMEIER, ORNL/USA
- Daniela KISELEV, Paul Scherrer Institute/Switzerland
- Hee-Seock LEE, PAL/Korea
- Sylvie LERAY, CEA-Saclay/France
- Tianjiao LIANG, Chinese Academy of Science/China
- Nikolai MOKHOV, Fermilab/USA
- Brad MICKLICH, ANL/USA
- Kazuyoshi MASAMOTO, KEK, Japan
- Michael PAUL, Hebrew University of Jerusalem/Israel
- Arjan PLOMPEN, European Commission Joint Research Center/Belgium
- Stefan ROESLER, CERN/Switzerland
- Yair YARIV, Soreq Nuclear Research Center/ Israel

Local Organizing Committee:

- Daniela ENE, ESS/Sweden
- Stuart ANSELL, ESS/Sweden
- Riccardo BEVILACQUA, ESS/Sweden
- Duy PHAN, ESS/Sweden
- Nina ÅSVATNE, ESS/Sweden

IOP Conf. Series: Journal of Physics: Conf. Series **1046** (2018) 011003 doi:10.1088/1742-6596/1046/1/011003

Participant List: ARIA17

Name	Institution	Position	City	Country
Dr. ADORISIO, Cristina	CERN		Geneva	Switzerland
Dr. ARAI, Masatoshi	European Spallation Source ERIC		Lund	Sweden
Dr. BATKOV, Konstantin	European Spallation Source ERIC		Lund	Sweden
Dr. BENTLEY, Phil	ESS		Lund	Switzerland
Dr. BEVILACQUA, Riccardo	European Spallation Source ERIC	Scientist	Lund	Sweden
Mr. BORGHI, Nicolò	European Spallation Source ERIC		Lund	Sweden
Dr. DAVID, Jean-Christophe	CEA		Gif sur Yvette	France
Dr. DI JULIO, Douglas	European Spallation Source ERIC		Lund	Sweden
DIAN, Eszter	HAS Centre for Energy Research		Budapest	Hungary
Dr. ENE, Daniela	European Spallation Source, ESS-ERIC		Lund	Sweden
Dr. FLEMING, Michael	UK Atomic Energy Authority		Abingdon, Oxon	United Kingdom
Dr. FROESCHL, Robert	CERN		Geneva	Switzerland
Dr. GEORGE, Kharashvili	Jefferson Lab		Newport News	United States
Dr. GEORGOBIANI, Dali	Facility for Rare Isotope Beams, Michigan State University	Radiation Transport Physicist	East Lansing	United States
HARTL, Monika	European Spallation Source ERIC		Lund	Sweden
Ms. ILIOPOULOU, Elpida	CERN		Geneva	Switzerland
Dr. JÄGERHOFER, Lukas	EBG MedAustron GmbH	Head of Radiation Protection Health and Safety	2700 Wiener Neustadt	Austria
Dr. KISELEV, Daniela	Paul Scherrer Institut		Villigen	Switzerland
KLINKBY, Esben	DTU		Roskilde	Denmark
Dr. KOSTIN, Mikhail	Facility for Rare Isotope Beams	Radiation Transport Staff Physicist	East Lansing	United States
Ms. LEE, Arim	Pohang Accelerator Laboratory	Researcher	Pohang	Korea, Republic of

IOP Conf. Series: Journal of Physics: Conf. Series **1046** (2018) 011003 doi:10.1088/1742-6596/1046/1/011003

Dr. LEE, Hee-Seock	Pohang Accelerator Laboratory / POSTECH		Pohang	Korea, Republic of
Dr. LEE, Sangjin	Institute for Basic Science (IBS)		Daejeon	Korea, Republic of
LEE, Yong Joong	European Spallation Source ERIC		Lund	Sweden
Dr. LIU, James	Stanford Linear Accelerator Center	Deputy Department Head of Radiation Protection Department	Menlo Park, CA	United States
Dr. MAGISTRIS, Matteo	CERN		Geneva	Switzerland
Dr. MARCHIX, Anthony	CEA Saclay		Saclay	France
Prof. MASUMOTO, Kazuyoshi	High Energy Accelerator Research Organization		Tsukuba	Japan
Dr. MATSUMURA, Hiroshi	High Energy Accelerator Research Organization (KEK)	Associate professor	Tsukuba, Ibaraki 305-0801	Japan
Dr. MOKHTARI ORANJ, Leila	Division of Advanced Nuclear Engineering/POSTECH		Pohang	Korea, Republic of
Dr. PERSSON, Per	European Spallation Source ERIC	Radioactive Waste Group Leader	Lund	Sweden
Mr. POLATO, Andrea	European Spallation Source ERIC		Malmö	Sweden
Mrs. POPOVA, Irina	ORNL		Oak ridge	United States
Dr. QUINTIERI, Lina	ENEA		Rome	Italy
Dr. RAKHNO, Igor	Fermi National Accelerator Laboratory		Batavia, IL 60510-0500	United States
ROESLER, Stefan	CERN		Geneva	Switzerland
SANTANA LEITNER, Mario	SLAC		Menlod Park	United States
Dr. SANTORO, Valentina	European Spallation Source ERIC		Lund	Sweden
Dr. SHETTY, Nikhil	ELI Beamlines		Prague	Czech Republic
Dr. SJÖSTRAND, Henrik	Uppsala Universitet		Uppsala	Sweden
Dr. SKORO, Goran	ISIS Neutron and Muon Source, Science		Didcot	United Kingdom

IOP Conf. Series: Journal of Physics: Conf. Series **1046** (2018) 011003 doi:10.1088/1742-6596/1046/1/011003

	and Technology Facilities Council			
Dr. STICHELBAUT, Frederic	IBA	Domain Expert	Louvain-la-Neuve	Belgium
Dr. SUBLET, Jean-Christophe	United Kingdom Atomic Energy Authority		Abingdon Oxfordshire OX14 3DB	United Kingdom
Dr. TCHELIDZE, Lali	ESS ERIC	Safety Group Leader, Accelerator Division	Lund	Sweden
Dr. THEIS, Christian	CERN		Geneva	Switzerland
Mr. TOYODA, Akihiro	High Energy Accelerator Reserch Organization		Tsukuba city Ibaraki prefecture	Japan
Mr. WEIXELBAUMER, Christoph	EBG MedAustron GmbH		Wiener Neustadt	Austria
Dr. WU, Qingbiao	Institute of High Energy Pyisics, Chinese Academy of Sciences		Dongguan	China
YU, Quanzhi	Institute of Physics, Chinese Academy of Scienses		Beijing	China
ZANINI, Luca	European Spallation Source ESS ERIC		Lund	Sweden

IOP Conf. Series: Journal of Physics: Conf. Series **1046** (2018) 011004 doi:10.1088/1742-6596/1046/1/011004

Peer review statement

All papers published in this volume of *Journal of Physics: Conference Series* have been peer reviewed through processes administered by the proceedings Editors. Reviews were conducted by expert referees to the professional and scientific standards expected of a proceedings journal published by IOP Publishing.

Nuclide production in spallation reactions: How useful are the simulations?

J-C David[1], A Boudard[1], J Cugnon[2], J Hirtz[1,3], S Leray[1], D Mancusi[4] and J-L Rodriguez-Sanchez[1]

[1] IRFU, CEA, Université Paris-Saclay, F-91191, Gif-sur-Yvette, France
[2] AGO department, University of Liège, allée du 6 août 17, bât. B5, B-4000 Liège 1, Belgium.
[3] Center for Space and Habitability, Universität Bern, CH-30 12 Bern, Switzerland
[4] DEN-Service d'étude des réacteurs et de mathématiques appliquées (SERMA), CEA, Université Paris-Saclay, F-91191, Gif-sur-Yvette, France.

E-mail: jean-christophe.david@cea.fr

Abstract. INCL (Liège IntraNuclear Cascade model) combined with a deexcitation code has been used a lot for numerous simulations of spallation reactions during the last two decades. We go back over some of those simulations to address the capabilities of such codes and to show some improvements. The four examples of simultation are: the EURISOL project, the MEGAPIE target, the ESS facility, and the cosmogenic nuclide production. The goal is to discuss respectively designing and optimisation, predictive power and reliability, feasibility and uncertainty estimate, and the use of modeling to mitigate lack of experimental data.

1. Introduction

Nuclide production simulations in spallation reactions are done for many years with dedicated codes. They are useful and even necessary sometimes when no experimental data are available. Four aspects are addressed in this proceedings and based on four different studies. The use of simulations to optimize the design of a facility is illustrated by the EURISOL-DS project carried out between 2005 and 2009. The reliability of the models is discussed using simulations done for the Megapie target irradiated in 2006. We tried to tackle the difficult question of uncertainty estimates within the ESS Technical Design Report (2010). Finally we show that simulations are required when no experimental data exist with the example of meteorite study.

Most of the calculations presented here are done with the intranuclear cascade model INCL combined to the deexcitation code Abla. It is also the opportunity to show some improvements done in those two codes during the last years, as well as the on-progress studies. However, the topics mentioned here are general and apply to all spallation models.

2. Optimisation: EURISOL-DS

The EURopean Isotope Separation On-Line radioactive ion beam facility Design Study (EURISOL-DS) was a project aiming at determining the main parameters of this future european facility dedicated to the study of exotic nuclei never studied before, while taking into account the two constraints: safety and cost. The isotopes are produced in two types of target: direct (primary proton beam induced spallation reactions) or fission (the primary beam delivers, from a converter target, a neutron flux, which induces fission in a surrounding uranium target). For our purpose we focus on the direct target.

Our goal was first to calculate the production rate of isotopes given by NuPPECC from targets using different materials and of different dimensions, with different energies for the primary proton beam, in order, in a second step, to define the best configurations. Five materials were chosen (an oxyde (Al2O3), a silicate (SiC), a molten salt (Pb), a refractory (Ta), and a heavy element to get heavy isotopes (UC3)). The targets were assumed to be cylinder and for each material four length and four radii were studied. The four possible energies were 0.5, 1, 1.5 and 2 GeV. Thus 320 configurations were run with two spallation models, both within MCNPX: INCL4.2-Abla ([1, 2]) and CEM2k [3] .

This method, associated to specific benchmarks of the models used, allowed to obtain the type of results displayed in Fig. 1, with the isotopes, the optimal material to produce them, the production rate given by the two codes and, not only the statistical error, but also, and surely more important, some comments on the results. Thus, for example, it was mentioned that, for the isotopes obtained after a long evaporation process, INCL4.2-Abla underestimated the rates whereas CEM2k overestimated them, or, that a specific benchmark should be done for light targets, when experimental data will be available, because results were different, with no way to disentangle between both models. Those remarks were done first to better define the reliability of the results and second to motivate future improvements in codes and/or to carry out new experiments.

Isotope	Optimal target(s) for $E = 1$ GeV	Maximal yield [atom/s] INCL4/ABLA	Maximal yield [atom/s] CEM2k	MCNP statistical error [%] INCL4/ABLA	MCNP statistical error [%] CEM2k	Additional remarks
^{9}Li	Al_2O_3; SiC	$2.5 \cdot 10^{11}$; $1.1 \cdot 10^{11}$	$4.5 \cdot 10^{11}$; $1.3 \cdot 10^{12}$	3; 4	2; 1	Remark (1)
^{11}Li	Al_2O_3	$1.8 \cdot 10^{9}$	$3.6 \cdot 10^{10}$	30	6	Remark (1)
^{7}Be	SiC	$1.2 \cdot 10^{13}$	$1.2 \cdot 10^{13}$	< 1	< 1	Remark (1)
^{11}Be	Al_2O_3	$1.3 \cdot 10^{11}$	$4.4 \cdot 10^{11}$	3	2	Remark (1)
^{12}Be	Al_2O_3	$3 \cdot 10^{10}$	$2.3 \cdot 10^{11}$	7	3	Remark (1)
^{17}Ne	SiC	$1.3 \cdot 10^{10}$	$3.7 \cdot 10^{9}$	30	20	Remarks (2), (3)
^{18}Ne	SiC	$1.9 \cdot 10^{12}$	$8.6 \cdot 10^{10}$	< 1	5	Remarks (2), (3)
^{25}Ne	Al_2O_3; Pb	$1.2 \cdot 10^{10}$; $3.7 \cdot 10^{8}$	$2.4 \cdot 10^{9}$; $3.9 \cdot 10^{9}$	13; 100	25; 30	Remark (3)
^{20}Mg	SiC	$2.3 \cdot 10^{11}$	$2.4 \cdot 10^{10}$	3	8	Remark (3)

Figure 1. Part of the results obtained during the EURISOL-DS project on the isotope production rates and their optimization in material and dimension. More details in [4].

Such a study is not achievable with no spallation models implemented in transport codes and the efficiency depends strongly on the reliability and quickness of computing, two often conflicting features.

3. Reliability: Megapie

More than the reliability by itself, it is what we can do when a tool is considered reliable enough that is adressed in this part with a result obtained within the Megapie project.

Megapie was a liquid Pb-Bi target prototype irradiated for four months in 2006 at the SINQ facility (PSI). Our goal was to estimate the nuclei produced in the target and in other parts (e.g., window) to assess the safety issues related to activation and volatile release. Regarding the latter point a dedicated experiment was carried out and performed at ISOLDE (Cern) to study the release of volatiles [5].

A cylindrical (L=20 cm, r=1 cm) Pb-Bi target was irradiated by a proton beam and mass distributions of numerous volatiles (He, Ne, Ar, Br, Kr, Cd, I, Xe, Hg, Po and At) were measured. This was also used to benchmark the calculation models. All elements were rather well reproduced, except one: astatine. No model was able to simulate the right shape of the mass distribution, with several order of magnitude for the light isotopes compared to the experimental data. Since some isotopes of astatine decay in Po and astatine is more volatile than Po, this element is important for safety reasons. Thus one tried to understand and solve the simulation problem.

Astatine from bismuth can be produced in two ways: i) the light isotopes are produced by interaction of the primary proton beam on bismuth with emission of neutrons and one and only one pion, and ii) the heavy isotopes are produced by a two-step process where helium nuclei, first emitted from primary reactions, induce astatine production on bismuth. At this time the INCL4.2 cascade was unable to emit helium and the Abla deexcitation code evaporated only alpha (no ^{3}He). New improved and extended versions were developed and then used, still within MCNPX. INCL4.6 [6] replaced INCL4.2 with emission of nucleons, pions and all light nuclei with mass below 8, with a nuclear potential for the pion and a special care for the low-energy reactions (case of the helium-induced reactions involved in the case of astatine). Abla was moved to Abla07 [7] with the possibility to emit all types of nuclei.

Those new versions were tested on the elementary processes playing a role in this study and the results were satisfactory, especially for the two-steps channel. However, the mass distribution remained badly reproduced and we had to make an assumption to be able to fit the experimental data. Since INCL4.6 combined to Abla07 was able to give good results regarding the elementary processes, the calculation results were probably not the ones to compare to the measurements. Noting that the measured production rates and the mass, so the half-life, of the isotopes were correlated, on the one hand, and, on the other hand, the calculations were usually done in the target because the release of the most common elements were instantaneous, one decided to add a release time to the astatine, for which no information were available. Doing so and testing several times, the models could fit the data with a time between 5h and 10h, as shown in Fig. 2

When a tool is reliable enough, as it is the case for some spallation models now, it can help determining parameters beyond its own domain. This has been demonstrated here, with the possible range of values for the astatine time release in a liquid lead-bismuth target, obtained thanks to a model benchmarked and validated in the domain of interest. Reliability is a goal to reach by the model developers and a necessary step to make the tool useful for a larger community.

4. Uncertainties: ESS

A calculation result given without its uncertainty has no meaning. Determining uncertainties is much more difficult than computing a result. That is probably the two reasons why spallation models are benchmarked before to be used. This empirical way defines the validity in a given domain.

Within the ESS Technical Design Report [9] the total activities in the spallation target had to be calculated and the related uncertainties estimated. This was the opportunity to try to go beyond the usual benchmarks. Calculations were done with two models, INCL4.6-Abla07 ([6, 7]) and CEM03 [10], implemented in MCNPX. A first, but crude, idea of the uncertainties was the comparison of both calculation results. If both models were basically identical, the uncertainties (differences) would come for coding. However, no model are identical and numerous verifications are done to strongly reduce this kind of uncertainty. The goal was then to find out the origins of uncertainties.

The activity comes from many isotopes, but fortunately only few of them are strong contributors (say higher than 1%). Thus uncertainties are supposed to be in the parameters

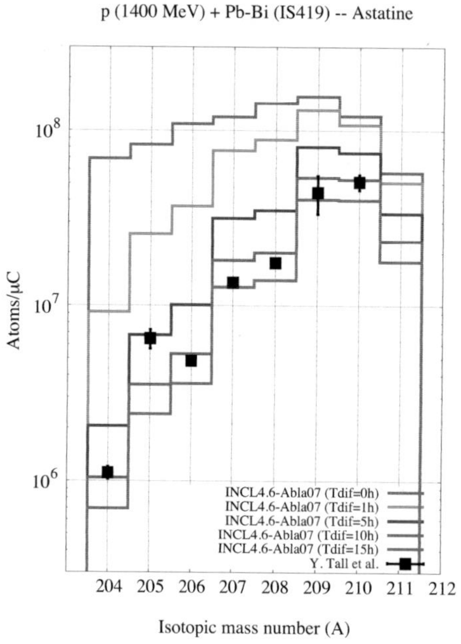

Figure 2. Mass distribution of the astatine isotopes released from a liquid Pb-Bi target irradiated by a 1.4 GeV proton beam. More details in [5] for the experiment and [8] for the modeling.

describing the contributor production. An isotope is produced by interaction of a particle with a target nucleus. Then, the particle flux and the elementary production cross section are the two parameters. Only neutron and proton play a significant role as projectile, the other particles (π, α, ...) can be neglected a priori, and the nucleon fluxes are known to be well simulated by the models. The uncertainties coming from the projectile fluxes are considered as negligible. Estimates of uncertainty of the cross sections are based on experimental data. A ratio is calculated between calculation and experiment and this ratio, depending on the projectile energy, is convolved with the particle spectrum. Several difficulties arise from this method. First, experimental data are not always available, and *similar* reactions must be sometimes used, second, data are sometimes not uniformly distributed in energy and interpolation and extrapolation are necessary, third, the data deal with proton almost exclusively. Concerning neutron, three domains of energy were defined. Below 25 MeV calculations were done with the evolution code CINDER [11] that uses a database for the cross sections, and so uncertainties were assumed to be negligible. Between 25 MeV and 200 MeV the experimental data are very scarce, which is an issue and reduces the validy of the method. Finally above 200 MeV neutron-induced and proton-induced cross sections are supposed to be the same. A schematic view of the method is given in Fig. 3

When applying carefully this complicated and tricky method, a correction of 12% was done on the target activity after nine years of cooling. All details can be read in [12] and [13]. However, having in mind the definitions taken from the GUM[14] of *uncertainty* (parameter, associated with the result of a measurement, that characterizes the dispersion of the values that could

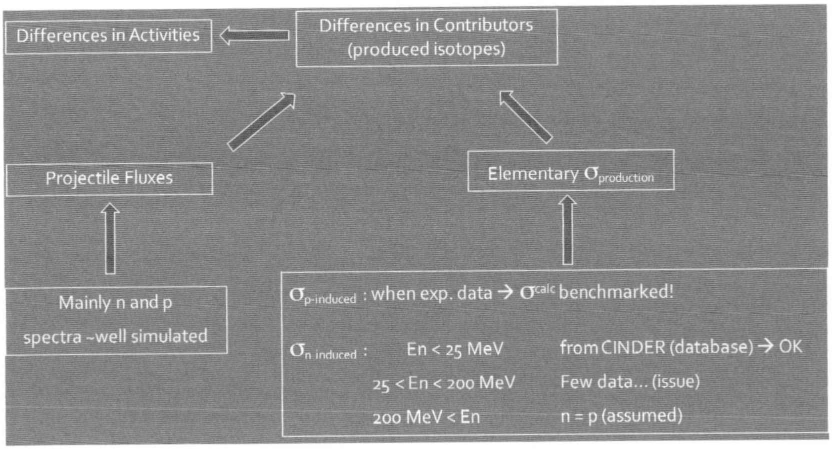

Figure 3. Summary of the quantities responsible of uncertainties in the ESS target activity.

reasonably be attributed to the measurand) and *error* (result of a measurement minus a true value of the measurand), it is clear that this method enables a correction of the calculation result based on experimental data, and so reduces the error, but does not give at all an estimate of the uncertainties. Actually uncertainty is related to precision while error is related to accurary. Both are important, useful and difficult to estimate, but they are assessed by different ways.

5. Lack of experimental data: meteorite study

It has been said in a previous section that the spallation models, when reliable enough, are useful to study other domains than nuclear reactions (it was material science and time release). In this section that is the use of spallation models as cross section generator which is addressed.

The study of meteorites can help to get information on the primordial solar system. Meteorites are interstellar bodies, originated from asteroides, and irradiated by cosmic rays during their travel. This irradiation produces cosmogenic nuclides and astrophysicists compared those measured production rates (PR) to calculated ones to obtain information on the history of the meteorite and/or the cosmic ray spectra as well. The equation used below exhibits the need of elementary nuclide cross sections.

$$PR = \sum_{i=1}^{N} c_i \frac{N_A}{A_i} \sum_k \int_0^\infty \sigma_{i,k}(E) J_k dE,$$

with i a target element, c_i its concentration and A_i its mass number, N_A the Avogadro's constant, k the type of particle inducing the reaction (primary and secondary particles), $\sigma_{i,k}(E)$ the production cross-section of the isotope by interaction of the particle k with the target i and with an incident energy E, and finally J_k the density flux of the particle k.

When available, the experimental cross sections are used, otherwise reaction models provide the missing data. Galactic cosmic rays are mainly made up protons (~87%) and helium (~12%) and their spectra are peaked in the spallation region that is roughly around 500 MeV, even if the range is very large. Then spallation models are good candidates to provide the needed cross sections. Three points can be mentioned. First, most of the experimental data deal with proton-induced reactions, but some models were significantly improved and they are now able to fit rather (sometimes very) well the data. The second point is about α, where the data are very scarce. Astrophysicists apply then a crude approach, i.e. they assume the result of an α-induced reaction as the sum of the result of two proton-induced reactions and to neutron-induced reaction, assuming that neutron and proton give the same result. Finally, the case of neutron-induced reactions. Those reactions are secondary reactions in the meteorites where numerous spallation neutrons are produced. Here again data are very scarce and we can mention the article of Leya and Michel [15], explaining how they extract those elementary cross sections from results on a thick target, where the proton contribution is removed and with the help of a guess function. Below are given three plots (Fig. 4(2) and 5(1)), one for each case, to show how INCL4.5[6]-Abla07 compares with the experimental data and the potential use of such models to mitigate the lack of data in studies like the meteorites.

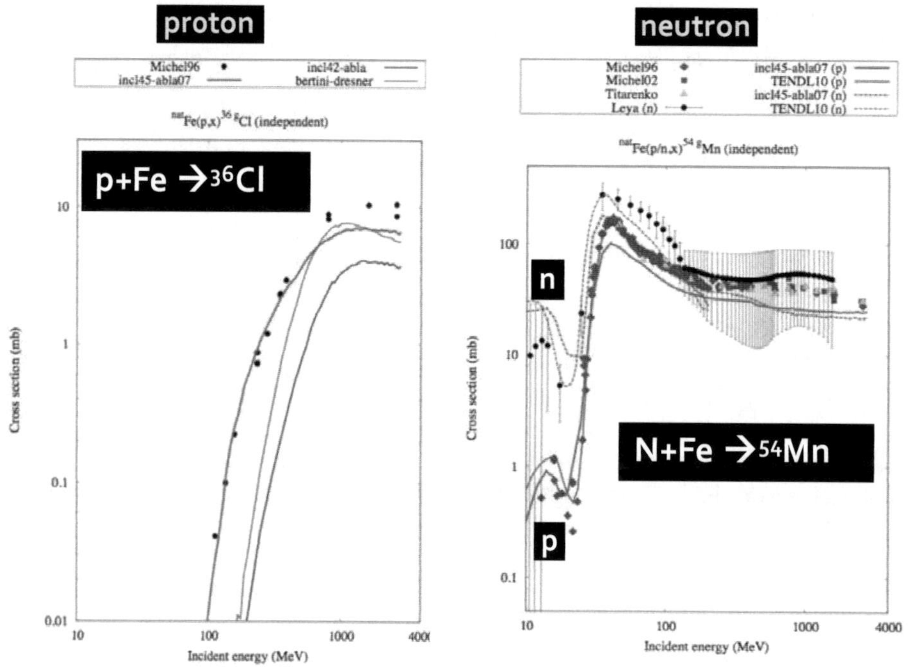

Figure 4. On the left, ^{36}Cl production cross sections from proton-induced reaction on natural Fe. More details on the figure and in [16] (page 100). On the right, ^{54}Mn production cross sections from neutron- and proton-induced reactions on natural Fe. More details on the figure and in [17].

6. Conclusion and Outlook

Nuclide production simulation in spallation reactions can be done by numerous models in several transport codes (for example see Table 1 in [13]). For the last twenty years significant improvements were done in this domain and some models can be considered particularly reliable. However, reliability, uncertainty, error are still difficult questions to answer clearly. This is the reason why benchmarks are even now needed. Some are done by the transport code developers (MCNPX, GEANT4, PHITS), some others performed in dedicated project ([18]), and sometimes must be done specifically.

Nevertheless, models are still improved and extended. For example, in the intranuclear cascade INCL, a specific study of few-nucleons removal has been initiated by D. Mancusi

IOP Conf. Series: Journal of Physics: Conf. Series **1046** (2018) 012001 doi:10.1088/1742-6596/1046/1/012001

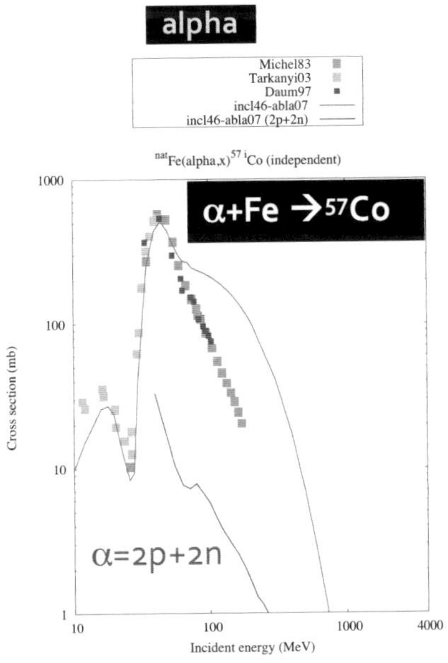

Figure 5. ^{57}Co production cross sections from α-induced reaction on natural Fe. The blue line is obtained by multiplying by four the result with a proton projectile (see text).

in [19] and is being generalized. Another extension is the implementation of strangeness (kaons, hyperons) to better describe the high-energy domain and extend the capabilities of INCL to kaon and hypernucleus production. These works are expected to be included in the next version of Geant4. The first results, compared to experimental data and other models, are encouraging.

References

[1] A. Boudard *et al.* 2002 *Phys. Rev.* C **66** 044615
[2] R. Junghans *et al.* 1998 *Nucl. Phys.* A **629** 635
[3] S. G. Mashnik and A. J. Sierk 2001 *Proc. AccApp00, November 12-16, 2000, Washington, DC (USA), ANS, La Grange Park, IL* 328-341; E-print: nucl-th/0011064
[4] S. Chabod *et al.* 2010 *Eur. Phys. J.* A **45** 131
[5] Y. Tall *et al.* 2008 *Volatile elements production rates in a proton-irradiated molten lead-bismuth target*, in Proceedings of the International Conference on Nuclear Data for Science and Technology, April 22-27, 2007, Nice, France, editors O.Bersillon, F.Gunsing, E.Bauge, R.Jacqmin, and S.Leray, EDP Sciences, 1069-1072 - DOI: 10.1051/ndata:07762
[6] A. Boudard *et al.* 2013 *Phys. Rev.* C **87** 014606
[7] A. Kelic, M.V. Ricciardi, and K.-H. Schmidt 2008 *Abla07 - towards a complete description of the decay channels of a nuclear system from spontaneous fission to multifragmentation* . Joint ICTP-IAEA Advanced Workshop on Model Codes for Spallation Reactions, 181222
[8] J.-C. David *et al.* 2013 *Eur. Phys. J.* A **49** 29
[9] ESS Technical Design Report; http://eval.esss.lu.se/cgibin/ public/DocDB/ShowDocument?docid=274
[10] S. G. Mashnik *et al.* 2008 *CEM03.03 and LAQGSM03.03 Event Generators for the MCNP6, MCNPX, and MARS15 Transport Codes* LANL Report LA-UR-08-2931; arXiv:0805.0751v2 [nucl-th]
[11] W. B. Wilson *et al.* 1998 *Status of CINDER'90 Codes and Data* LA-UR-98-361
[12] A. Leprince *et al.*2014 *Reliability and use of the INCL4.6- Abla07 spallation model in the frame of the European Spallation Source (ESS) target design* CEA Saclay Internal report: IRFU-14-25
[13] J.-C. David 2015 Eur. *Phys. J.* A **51** 68
[14] GUM (Guide to the expression of Uncertainty in Measurement), JCGM 100:2008; http://www.bipm.org/fr/publications/guides/gum.html

[15] I. Leya and R. Michel 2011 *Nucl. Instrum. Meth.* B **269** 2487
[16] J.-C. David 2012 *Spallation: comprendre ou prédire ! ?* Report of Habilitation à Diriger des Recherches (in french), Irfu-12-259; http://tel.archives-ouvertes.fr/tel-00811587
[17] J.-C. David *et al.* 2011 *Memorie della Società Astronomica Italiana,* **82** 909
[18] S. Leray *et al.* 2011 *J. Korean Phys. Soc.* **59** 791; doi:10.3938/jkps.59.791
[19] D. Mancusi *et al.* 2015 *Phys. Rev.* C **91** 034602

IOP Conf. Series: Journal of Physics: Conf. Series **1046** (2018) 012002 doi:10.1088/1742-6596/1046/1/012002

High-Energy Activation Simulation Coupling TENDL and SPACS with FISPACT-II

Michael Fleming, Jean-Christophe Sublet and Mark Gilbert

UK Atomic Energy Authority, Culham Science Centre, Abingdon OX14 3DB, United Kingdom

E-mail: michael.fleming@ukaea.uk

Abstract. To address the needs of activation-transmutation simulation in incident-particle fields with energies above a few hundred MeV, the FISPACT-II code has been extended to splice TENDL standard ENDF-6 nuclear data with extended nuclear data forms. The JENDL-2007/HE and HEAD-2009 libraries were processed for FISPACT-II and used to demonstrate the capabilities of the new code version. Tests of the libraries and comparisons against both experimental yield data and the most recent intra-nuclear cascade model results demonstrate that there is need for improved nuclear data libraries up to and above 1 GeV. Simulations on lead targets show that important radionuclides, such as ^{148}Gd, can vary by more than an order of magnitude where more advanced models find agreement within the experimental uncertainties.

1. Introduction

The simulation of activation-transmutation processes in nuclear environments requires knowledge of all reaction cross sections that are energetically allowable. This is often handled through the use of nuclear data libraries, which contain cross sections for every unique channel, for example, with incident neutrons, (n,p), (n,2n), (n,α), etc. Below 30 MeV, the number of reaction channels are sufficiently small, and the channels sufficiently distinguishable, that this method has remained in standard use. However, as the incident energy climbs, the number of channels rapidly expands and nuclear data libraries utilise a residual product based data structure which can accommodate arbitrarily many products.

The TENDL-2015 [1, 2] neutron and charged-particle nuclear data files cover some 2809 targets with complete reaction data including all emitted particles for incident energies up to 200 MeV. These data are contained within the international standard ENDF-6 format [3], with full product energy-angle data (in `MF=6`) that may be processed into production cross-section data (in `MF=10`) and utilised by codes such as FISPACT-II [4, 5] for activation simulations. The driving engine behind TENDL, the TALYS code [6], performs model calculations to predict the cross sections for each of these yields. This incurs significant computational expense and, even with the most recent, advanced data, does not provide data for the full spallation yield 'tail' including nucleon differences in excess of a few dozen. This is analogous to the case of fission, where each individual product from the fission process is not considered, but a separate data set of fission yields is coupled with the total fission cross section to determine creation rates for every product. It is standard practice to use other models and software (e.g. GEF [7]) to complement the data.

At energies of a few tens or hundreds of MeV and higher, intra-nuclear cascade models are commonly utilised to simulate the reaction process, generating a set of directly emitted particles and residuals that undergo evaporation, fission and fragmentation processes. Many models (see [8] for a survey and references therein) have been engineered to be used in Monte-Carlo transport simulations, providing an intra-nuclear Monte-Carlo process within the inter-nuclear transport calculation. This method has been demonstrated to be highly effective for transport calculations, but all outputs must be converged in these Monte-Carlo simulations. While for low-mass particle transport this is often a reasonable computational problem, for the production of *all residual nuclei*, it remains a challenge. Individual products of radiological importance may be produced in only one of several thousand or fewer events, requiring (potentially many) millions of reaction simulations in every region of interest.

While in the well-studied fission and fusion applications the nuclear reactions above 30 MeV do not play an essential role, in a large number of related research, such as materials damage studies, nuclear medicine, accelerator-driven research or space exploration, the nuclide inventories and resulting observables from such high-energy reactions are of significant importance.

2. Activation-transmutation simulation

While in some simulation systems (c.f. [9]) the record of each residual nuclide is retained for post-processing and activation-transmutation calculation, an alternative is to rely completely upon nuclear data libraries for this secondary calculation. With this alternative, the incident particle fluxes and spectra are calculated in each region and coupled with complete nuclear data for all relevant energies and reaction channels. This is shown schematically in Figure 1. The 'low-energy' nuclear data referenced in Figure 1 may range from 10^{-5} eV to 30 MeV or run as high as 200 MeV, but the principle is the same: results from Monte-Carlo intra-nuclear model calculations are used to determine *reaction rates* for radionuclide production. 'All-energy' nuclear data is required to compensate for the lack of residual tally data, providing reaction rates for all products.

The FISPACT-II implementation[1] requires the standard 'low-energy' ENDF-6 nuclear data libraries, as well as an extended data library of residual production `MF=10` format. The use of a new keyword, **USESPALLATION**, causes data from both of these sources to be spliced in the reading of energy-dependent cross sections. The user supplies two keyword options: an energy cutoff (in MeV) E_{cut} and a number of isotopes N_{iso} that this should be treated. This is followed by a list of isotopes, typically the stable isotopes present in some irradiated material. Inputting $N_{iso} = -1$ causes FISPACT-II to splice data for all isotopes that possess data. The one group cross sections are then calculated as;

$$\overline{\sigma} = \frac{\left[\sum_{\{i|E_i<E_{cut}\}} \phi_i \, \sigma_{LE,i}\right] + \left[\sum_{\{j|E_j\geq E_{cut}\}} \phi_j \, \sigma_{HE,j}\right]}{\sum_k \phi_k}, \quad (1)$$

where the low- and high-energy library cross sections are subscripted as *LE* and *HE*, respectively.

Once the rate equation matrix elements have been calculated, the full machinery of FISPACT-II may be used to calculate all nuclide inventories and responses, as shown in Figure 2 for a one year, 1 GeV proton irradiation of natural tungsten. The production of each nuclide is followed by complete tracking of all decay processes, showing the relative importance of all possible radionuclides. In this case, the presence of ^{148}Gd, with a half-life of approximately 75 years, will remain a dominant contributor and, as an α-emitting nuclide, one of radiological importance [10].

[1] As of `FISPACT-II` v4.00

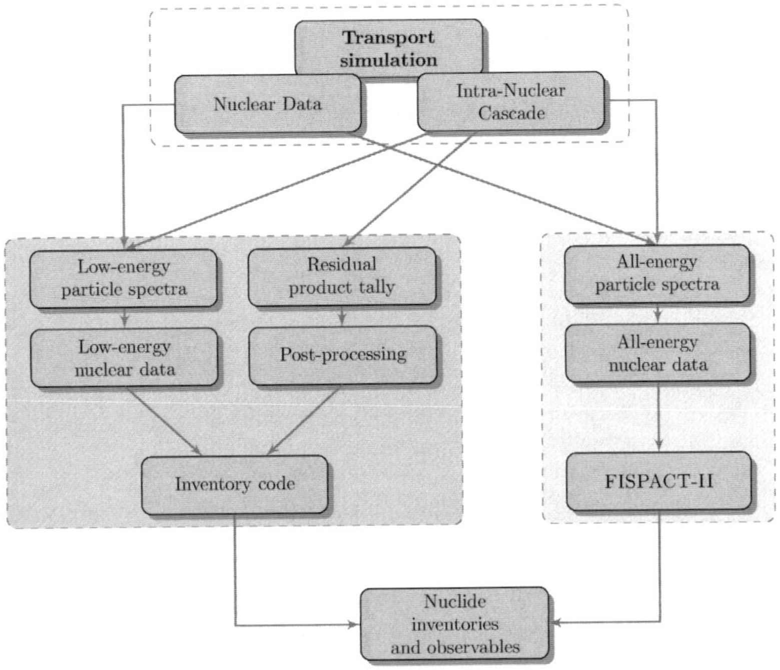

Figure 1. Schematic for activation-transmutation inventory calculations, using a Monte-Carlo transport simulation (yellow), coupled with either (red) separate nuclear data and residual history data or (green) pure nuclear data library calculations.

3. High-energy nuclear data

The essential ingredient required for the full multi-group nuclear data approach is a complete nuclear data library, including all reaction channels and products. The only non-deprecated nuclear data libraries known to the authors are the JENDL-2007/HE [12] and HEAD-2009 [13] libraries, whose content is briefly summarised in Table 1. The HEAD-2009 data is a high-energy extension to complement other nuclear data below 150 MeV and has been created using a suite of model codes, as detailed in the reference report. These contain full emitted particle energy spectra in `MF=6`, but have been processed into residual production group-wise data for FISPACT-II. The JENDL-2007/HE data is a high-energy extension of JENDL data files that utilise different file formats depending on the evaluator methodology.

Table 1. Summary of high-energy nuclear data libraries.

Library	Targets	Energy range (MeV)
HEAD-2009 (neutron/proton)	H1-Bi210 (629 files)	150-1000
JENDL-2007/HE (neutron/proton)	H1-Am242m (106 files)	up to 3000

While the JENDL-2007/HE files only cover 106 targets, the graphs of the yield data in Figures 3 and 4 demonstrate that they are more complete and cover all products. The mixed choice between these two libraries are between target coverage and product coverage. However,

IOP Conf. Series: Journal of Physics: Conf. Series **1046** (2018) 012002 doi:10.1088/1742-6596/1046/1/012002

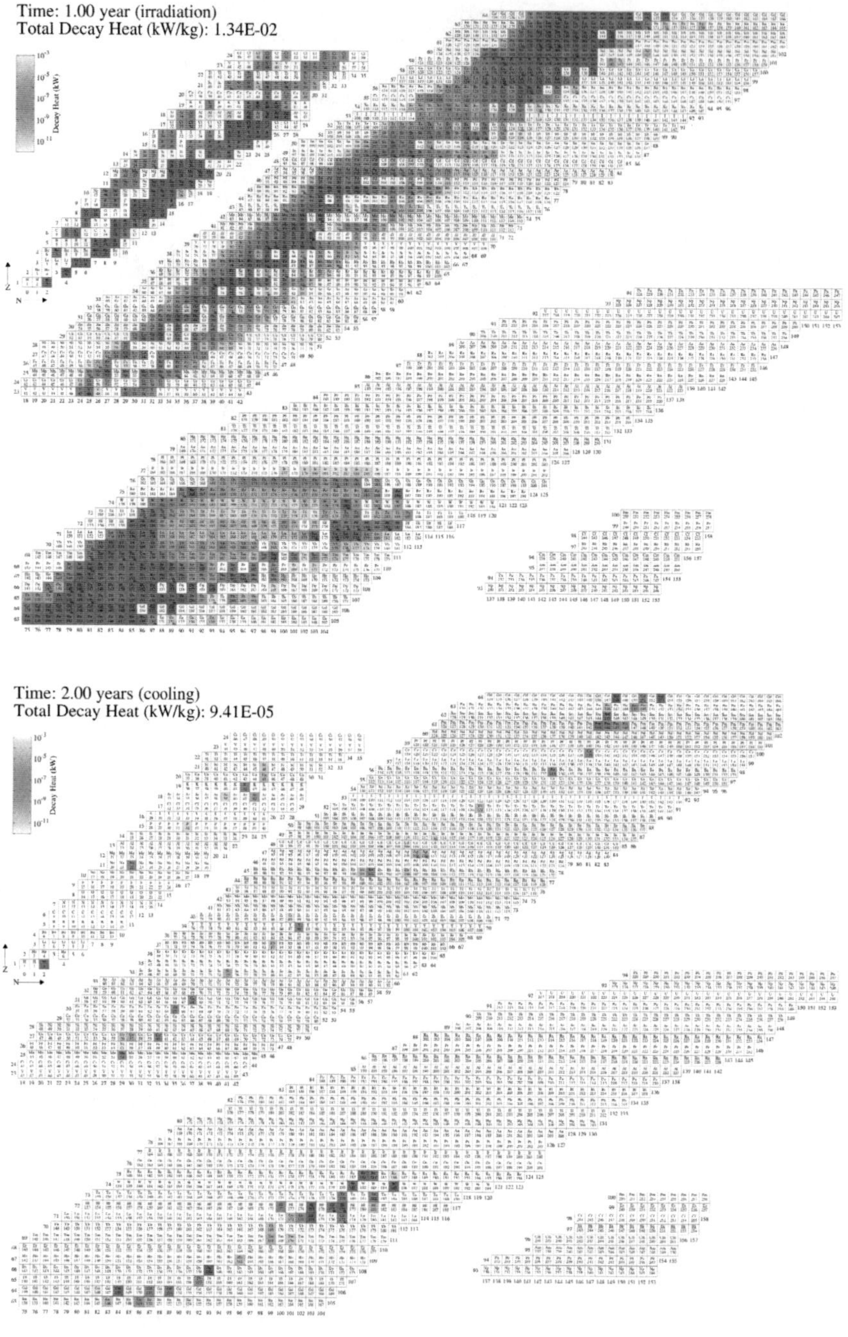

Figure 2. FISPACT-II decay heat simulation [11] for a one year, 1 GeV proton irradiation of natural tungsten, using spliced data from TENDL-2015 up to 150 MeV and the HEAD-2009 library to 1 GeV, supplemented with SPACS data. The top figure shows the decay heat immediately at the end of irradiation and the bottom figure after two years of cooling. The color axis provides the decay heat response for each radionuclide, ranging from tritium to tungsten and above.

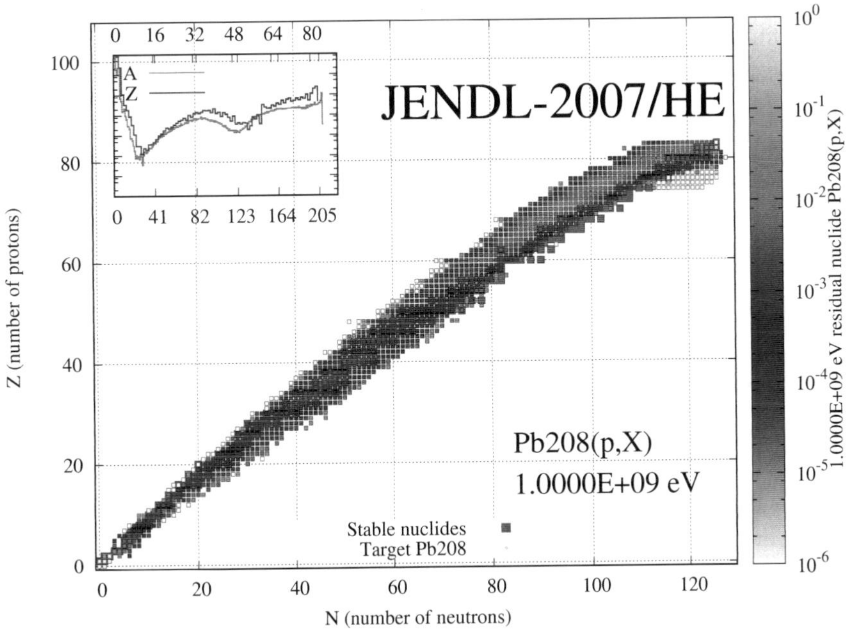

Figure 3. JENDL-2007/HE 1 GeV proton-induced residual product yields, showing a complete data set from target to proton.

the production of more than 1000 products per target highlights the need for more than simply 629 target files, since each long-lived product will itself require negative rate equation matrix elements. To accommodate a true inventory calculation for high-energy reactions, a complete set of reactions are required, including the radionuclide products.

To generate the full tabulated data required for all targets in activation simulations, the semi-empirical isotopic spallation simulation code SPACS [14] was utilised to build complete isotopic yield data files for incident neutrons and protons with energies from 100 MeV to a few GeV. These can be read directly into the FISPACT-II code and used with the complete nuclear reaction data of TENDL to provide a consistent simulation framework including high-energy activation residual products down to approximately half of the target mass.

Moving beyond the completeness of the nuclear data files, the quality should be determined by comparison with available experimental data. The 1 GeV proton-induced Pb208 data of Enqvist et al [15] provides one such comparison, where the agreement found between standard physics list options within Geant4 [16] and the experimental data is much greater than with the nuclear data libraries, as shown in Figure 5.

Of particular interest, given the radiological importance of ^{148}Gd, is the mass yield for 148 as contained within the nuclear data libraries and as predicted by model codes. The order-of-magnitude errors presented by the available nuclear data files is not surprising, given the relative spread of simulation results, but underlines the potential for improvement. As Figure 5 demonstrates, INCL-based calculations and SPACS provide significant improvement in the calculation of this mass yield, as well as many others.

4. Discussion

The recent modifications to FISPACT-II allow users to employ nuclear data libraries to cover both the energies up to 200 MeV with TENDL, as well as extended nuclear data forms up to

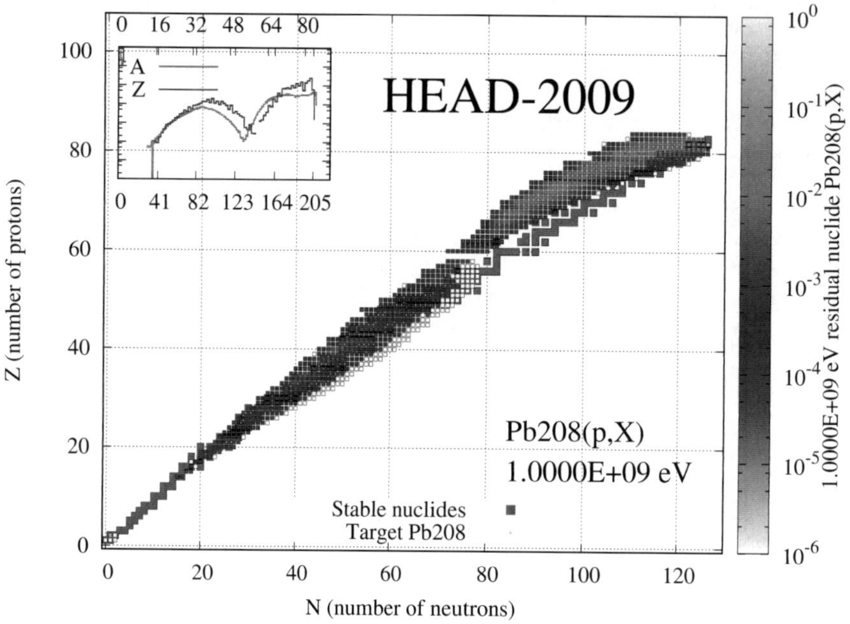

Figure 4. HEAD-2009 1 GeV proton-induced residual product yields, showing an incomplete data set from target to proton, but offering a larger number of incident targets with one reproducible evaluation methodology.

several GeV. For high-energy activation-transmutation calculations, this offers the significant advantage of using converged, reproducible residual nuclide production cross sections, rather than attempting to converge spallation residual tallies.

The remaining challenge is the production of accurate and reliable nuclear data files, although this is identical to the challenge for model codes within Monte-Carlo transport, since the same models are utilised in the production of nuclear data libraries. A survey of the available nuclear data files, including HEAD-2009 and JENDL-2007/HE, has found that both the targets and products are not fully covered in the data. The use of deprecated of intra-nuclear cascade and de-excitation models has been demonstrated with comparisons using the most recent models as available in Geant4. Systematic data generated with SPACS can offer yields for targets not considered in HEAD-2009 or JENDL-2007/HE, but without the processes for and products from fission and fragmentation.

New nuclear data libraries employing the best model codes, as employed in standard Monte-Carlo transport simulations, are required to provide the most accurate activation-transmutation simulations. To ensure reliability, traceability and reproducibility of the nuclear data libraries, precise model versions must be employed in a automated process, much as the TENDL nuclear data libraries are generated using a robust and repeatable method [19, 1].

Acknowledgements

This work was funded by the Research Councils United Kingdom Energy Programme under grant EP/P012450/1. M Fleming would like to thank J-Ch. David, D. Mancusi, A. Koning and K-H. Schmidt for helpful conversations and insights.

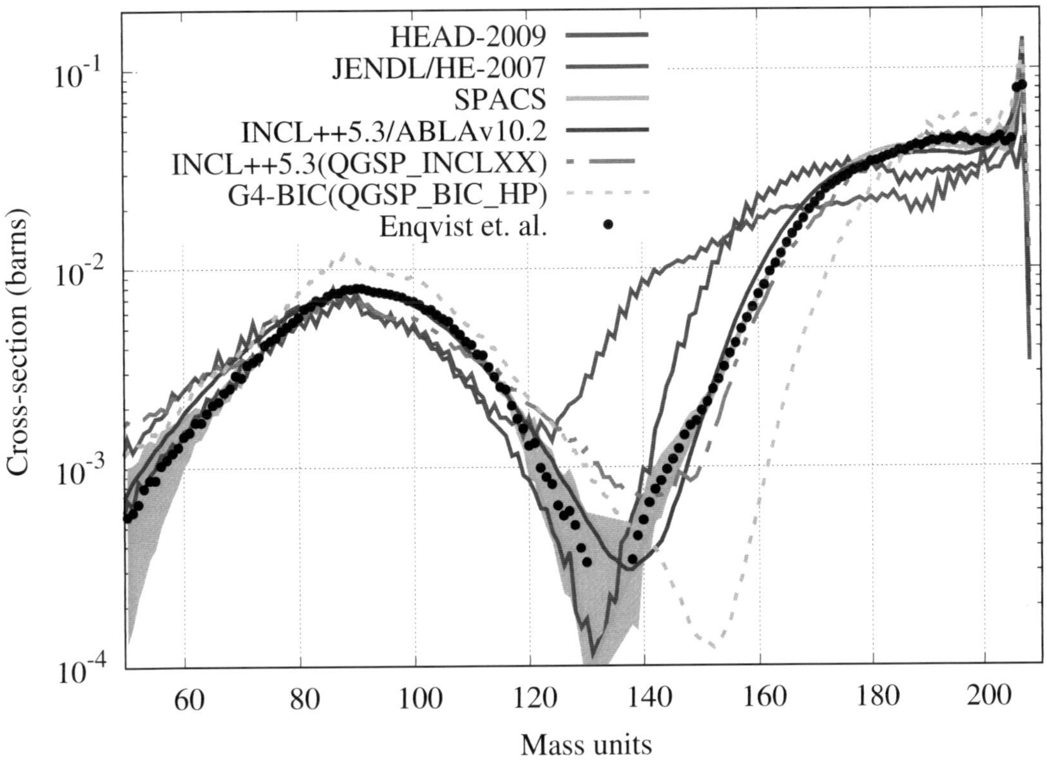

Figure 5. Comparison of nuclear data libraries, SPACS and intra-nuclear cascade model results for 1 GeV proton-induced Pb208 residual product cross sections, compared with experimental data from Enqvist et al [15] with experimental uncertainties in grey. Geant4 simulations were used with multiple physics lists and using an ABLA [17] optional de-excitation mode coupled with INCL++ [18].

References

[1] A. Koning and D. Rochman, "Modern Nuclear Data Evaluation with the TALYS Code System," *Nuclear Data Sheets*, vol. 113, no. 12, pp. 2841 – 2934, 2012. Special Issue on Nuclear Reaction Data.

[2] A. Koning, D. Rochman, S. Van der Marck, J. Kopecky, J.-C. Sublet, M. Fleming, S. Pomp, H. Sjostrand, R. Forrest, E. Bauge, and H. Henriksson, "TENDL-2015: TALYS-based Evaluated Nuclear Data Library." https://tendl.web.psi.ch/tendl_2015/tendl2015.html, 2015.

[3] M. Herman and A. Trkov, eds., *ENDF-6 Formats Manual, Data Formats and Procedures for the Evaluated Nuclear Data File ENDF/B-VI and ENDF/B-VII*, vol. BNL-90365-2009 Rev. 2. Brookhaven National Laboratory, Nov. 2011.

[4] J.-C. Sublet, J. Eastwood, J. Morgan, M. Gilbert, M. Fleming, and W. Arter, "Fispact-ii: An advanced simulation system for activation, transmutation and material modelling," *Nuclear Data Sheets*, vol. 139, no. Supplement C, pp. 77 – 137, 2017. Special Issue on Nuclear Reaction Data.

[5] J.-Ch. Sublet, J. Eastwood, J. Morgan, M. Fleming, and M. Gilbert, "FISPACT-II User Manual," Tech. Rep. UKAEA-R(11)11 Issue 7, UKAEA, Sept. 2015.

[6] A. Koning, H. S., and D. M., "TALYS-1.8 A nuclear reaction program." http://www.talys.eu/, 2017. NRG Nuclear Research and Consultancy Group.

[7] K.-H. Schmidt, B. Jurado, C. Amouroux, and C. Schmitt, "Special Issue on Nuclear Reaction Data General Description of Fission Observables: GEF Model Code," *Nuclear Data Sheets*, vol. 131, pp. 107 – 221, 2016.

[8] J. C. David, "Spallation reactions: A successful interplay between modeling and applications," *The European Physical Journal A*, vol. 51, p. 68, Jun 2015.

[9] C. Petrovich, "SP-FISPACT2001 A Computer Code for Activation and Decay Calculations for Intermediate Energyes. A Connection of FISPACT with MCNPX," Tech. Rep. RT/ERG/2001/10, 2001.

[10] K. Kelley, N. Hertel, E. Pitcher, M. Devlin, and S. Mashnik, "148Gd production cross section measurements for 600- and 800-MeV protons on tantalum, tungsten, and gold," *Nuclear Physics A*, vol. 760, no. 3, pp. 225 – 233, 2005.

[11] M. R. Gilbert, L. W. Packer, J.-C. Sublet, and R. A. Forrest, "Inventory Simulations under Neutron Irradiation: Visualization Techniques as an Aid to Materials Design," *Nuclear Science and Engineering*, vol. 177, no. 3, pp. 291–306, 2014.

[12] Y. Watanabe, "Status of JENDL High Energy File," *Journal of the Korean Physical Society*, vol. 59, no. 2, pp. 1040–1045, 2011.

[13] Y. Korovin, A. Natalenko, A. Stankovskiy, S. Mashnik, and A. Konobeyev, "High energy activation data library (HEAD-2009)," *Nuclear Instruments and Methods in Physics Research Section A: Accelerators, Spectrometers, Detectors and Associated Equipment*, vol. 624, no. 1, pp. 20 – 26, 2010.

[14] C. Schmitt, K.-H. Schmidt, and A. Kelić-Heil, "SPACS: A semi-empirical parameterization for isotopic spallation cross sections," *Phys. Rev. C*, vol. 90, p. 064605, Dec 2014.

[15] T. Enqvist, W. Wlazlo, P. Armbruster, J. Benlliure, M. Bernas, A. Boudard, S. Czajkowski, R. Legrain, S. Leray, B. Mustapha, M. Pravikoff, F. Rejmund, K.-H. Schmidt, C. Stéphan, J. Taieb, L. Tassan-Got, and C. Volant, "Isotopic yields and kinetic energies of primary residues in 1 A GeV 208Pb+p reactions," *Nuclear Physics A*, vol. 686, no. 1, pp. 481 – 524, 2001.

[16] S. Agostinelli, J. Allison, K. Amako, J. Apostolakis, H. Araujo, P. Arce, M. Asai, D. Axen, S. Banerjee, G. Barrand, F. Behner, L. Bellagamba, J. Boudreau, L. Broglia, A. Brunengo, H. Burkhardt, S. Chauvie, J. Chuma, R. Chytracek, G. Cooperman, G. Cosmo, P. Degtyarenko, A. Dell'Acqua, G. Depaola, D. Dietrich, R. Enami, A. Feliciello, C. Ferguson, H. Fesefeldt, G. Folger, F. Foppiano, A. Forti, S. Garelli, S. Giani, R. Giannitrapani, D. Gibin, J. G. Cadenas, I. Gonzlez, G. G. Abril, G. Greeniaus, W. Greiner, V. Grichine, A. Grossheim, S. Guatelli, P. Gumplinger, R. Hamatsu, K. Hashimoto, H. Hasui, A. Heikkinen, A. Howard, V. Ivanchenko, A. Johnson, F. Jones, J. Kallenbach, N. Kanaya, M. Kawabata, Y. Kawabata, M. Kawaguti, S. Kelner, P. Kent, A. Kimura, T. Kodama, R. Kokoulin, M. Kossov, H. Kurashige, E. Lamanna, T. Lampn, V. Lara, V. Lefebure, F. Lei, M. Liendl, W. Lockman, F. Longo, S. Magni, M. Maire, E. Medernach, K. Minamimoto, P. M. de Freitas, Y. Morita, K. Murakami, M. Nagamatu, R. Nartallo, P. Nieminen, T. Nishimura, K. Ohtsubo, M. Okamura, S. O'Neale, Y. Oohata, K. Paech, J. Perl, A. Pfeiffer, M. Pia, F. Ranjard, A. Rybin, S. Sadilov, E. D. Salvo, G. Santin, T. Sasaki, N. Savvas, Y. Sawada, S. Scherer, S. Sei, V. Sirotenko, D. Smith, N. Starkov, H. Stoecker, J. Sulkimo, M. Takahata, S. Tanaka, E. Tcherniaev, E. S. Tehrani, M. Tropeano, P. Truscott, H. Uno, L. Urban, P. Urban, M. Verderi, A. Walkden, W. Wander, H. Weber, J. Wellisch, T. Wenaus, D. Williams, D. Wright, T. Yamada, H. Yoshida, and D. Zschiesche, "Geant4 - a simulation toolkit," *Nuclear Instruments and Methods in Physics Research Section A: Accelerators, Spectrometers, Detectors and Associated Equipment*, vol. 506, no. 3, pp. 250 – 303, 2003.

[17] A. Heikkinen, P. Kaitaniemi, and A. Boudard, "Implementation of INCL cascade and ABLA evaporation codes in Geant4," *Journal of Physics: Conference Series*, vol. 119, no. 3, p. 032024, 2008.

[18] S. Leray, D. Mancusi, P. Kaitaniemi, J. C. David, A. Boudard, B. Braunn, and J. Cugnon, "Extension of the Liege Intra Nuclear Cascade model to light ion-induced collisions for medical and space applications," *Journal of Physics: Conference Series*, vol. 420, no. 1, p. 012065, 2013.

[19] A. Koning and D. Rochman, "Towards sustainable nuclear energy: Putting nuclear physics to work," *Annals of Nuclear Energy*, vol. 35, no. 11, pp. 2024 – 2030, 2008.

IOP Conf. Series: Journal of Physics: Conf. Series **1046** (2018) 012003 doi:10.1088/1742-6596/1046/1/012003

Proton induced activity in graphite - comparison between measurement and simulation

Daniela Kiselev, Ryan Bergmann, Dorothea Schumann, Vadim Talanov and Michael Wohlmuther

Paul Scherrer Institut, 5232 Villigen PSI, Switzerland

E-mail: Daniela.Kiselev@psi.ch

Abstract. The Paul Scherrer Institut (PSI) operates the Meson production target stations E and M with 590 MeV protons at currents of up to 2.4 mA. Both targets consist of polycrystalline graphite and rotate with 1 Hz due to the high power deposition (40 kW at 2 mA) in Target E. The graphite wheel is regularly exchanged and disposed as radioactive waste after a maximum of 3 to 4 years in operation, which corresponds to about 30 to 40 Ah of proton fluence. For disposal, the nuclide inventory of the long-lived isotopes ($T_{1/2} > 60$ d) has to be calculated and reported to the authorities. Measurements of gamma emitters, as well as ^{3}H, ^{10}Be and ^{14}C, were carried out using different techniques. The measured specific activities are compared to Monte Carlo particle transport simulations performed with MCNPX2.7.0 using the BERTINI-DRESNER-RAL (default model in MCNPX2.7.0) and INCL4.6/ABLA07 as nuclear reaction models.

1. Motivation and introduction

When radioactive material has to be disposed of, the authorities in Switzerland require a complete nuclide inventory for long-lived isotopes ($T_{1/2} > 60$ d). A complete nuclide inventory can only be achieved with reasonable effort via simulation, and for important nuclei, these calculations have to be validated. Often measuring important isotopes is not straight-forward and requires a chemical treatment beforehand. The authorities are not only interested in the radioisotopes with the highest activities or dose rates, but also in isotopes which are particularly problematic for the environment, such as α-emitters, volatile or outgassing substances, and elements known for migration or chemical reactions in final repositories.

Under irradiation of 590 MeV protons, mainly ^{3}H, ^{7}Be, ^{10}Be and ^{14}C are produced in pure graphite. ^{7}Be has a short life time of only 53 days, and it is usually already completely decayed when the graphite targets are ready for disposal. However, the activity of ^{7}Be determines the high dose rate of a few Sv/h at the graphite wheel a few days after operation in-beam. Contrary to it, ^{10}Be and ^{14}C are very long-lived. Carbon in an oxidized gaseous state is a candidate for migration out of the final repository into the biosphere and geosphere. Therefore, when disposing radioactive graphite, special measures are taken which are known from decommissioning of nuclear reactors. Tritium with a half-life of 12.3 a is produced in large quantities due to spallation (fragmentation of nucleus). This isotope is known for later outgassing from components. Since graphite is a porous material, it is expected that more tritium is released in graphite than from widely used metals like steel and copper. However, due to its low energy β-particle, tritium is not very harmful. It is also released in large quantities from nuclear power plants.

In addition, graphite used as moderator sticks in nuclear reactors is usually accompanied by large quantities of ^{36}Cl, a long-lived radioisotope, which is also a candidate for migration and outgassing.

In this report, a comparison is made of the calculated and measured specific activities of very pure graphite irradiated by a 590 MeV proton beam. The comparison is done on the following radioisotopes: ^{3}H, ^{7}Be, ^{10}Be, ^{14}C, ^{22}Na, ^{26}Al, ^{44}Ti, ^{54}Mn. Isotopes with mass equal to and lower than carbon are produced directly from carbon, whereas the heavier isotopes are produced from impurities. Besides the uncertainties in the impurities, this simple system is well suited for benchmarking. In addition, the long standing puzzle to reproduce the large activity of ^{14}C measured in [1, 2] by LAHET [3] or later MCNPX2.5.0 [4] calculations [5] makes this system particularly interesting. The experimental values in [1] were later reproduced by [6]. Two nuclear reaction models are used for comparison: The default model in MCNPX2.5.0 [4] and MCNPX2.7.0 [7], the BERTINI-DRESNER-RAL [8–11], as well as the improved INCL4.6/ABLA07 [12, 13].

2. Meson production target station E

At PSI there are two meson production target stations, called M and E. Both consist of a graphite target and deliver not only mesons, i.e. mainly pions, which later decay to muons. Pions and muons are used for research in particle physics and for muon spin-resonance applications (μSR).

The meson target stations are driven by 590 MeV protons with beam currents of up to 2.4 mA from the High Intensity Proton Accelerator (HIPA), which delivers the most intense continuous proton beam in the world. The protons are produced in a compact electron cyclotron resonance (ECR) source [14] and gain 870 keV by a Crockcroft-Walton generator. Afterwards they are further accelerated in two cyclotrons called "Injector2" and "Ring". The protons leave the Injector2 with an energy of 72 MeV and are transferred to the Ring, where they are accelerated to their final energy of 590 MeV. The beam is transferred to the first meson production target station, Target M, with an effective graphite thickness of 5 mm, or optionally kicked to the ultra-cold neutron source (UCN) for a few seconds. About 18 m later, the beam passes through the meson production target station, Target E, with an variable thickness of 40 mm or 60 mm. In Target E region, the protons lose 15 MeV of their initial energy and about 30 % of their intensity. The losses are caused mainly due to multiple scattering in the graphite of Target E, which also leads to additional divergence of the beam. For further transport, the beam is shaped by a subsequent collimator system, which is responsible of 2/3 of the beam losses. The beam loses at the Target M region are only 1 %. After the meson production targets, the beam is bent downwards then upwards to irradiate the spallation neutron source SINQ from beneath. At SINQ, thermal and cold neutrons are produced for material structure research and radiography from a zircalloy-clad lead target. If SINQ is not ready to receive beam, protons are instead steered to the beam dump after the meson production targets, keeping Target E and Target M in operation.

Only samples from Target E are investigated in this report. The insert of Target E is shown in figure 1. Two inserts exist since the bearing might have to be changed in one operation period. Two challenges for Target E are the high radiation levels as well as the high power deposition. The 590 MeV protons of 2.4 mA deposit about 50 kW in the 40 mm thick graphite of Target E and even more in the 60 mm target. To reduce the beam spread, a light material was chosen for the target material. Beryllium could not sustain the radiation dose, but polycrystalline graphite could. To dissipate the heat, the target was designed as a wheel rotating at 1 Hz. In the vicinity of the target, water-cooled copper removes the radiated heat. The graphite is heated to about 1500 ^{o} C at the beam spot. As graphite is an anisotropic material by nature, the thermal expansion coefficients are dependent on the lattice axis. This leads to extra thermal stress and finally to deformation or even cracks. Therefore, polycrystalline graphite, which consists of a

Figure 1. Insert of target E. The direction of the beam is also indicated.

large number of small single crystallites irregularly arranged in space, is used. This results in almost isotropic physical properties.

2.1. Description of sample Targets

The three targets, called E89, E83 and E92, which were used for this analysis were all made from SGL-CARBON graphite but from different grades. The target E79 uses EK94, E83 was produced from R6400, and E92 is made of R6510. The later grades have improved isotropic properties. They have smaller grains which lead to less porosity, slightly larger density, and better radiation resistance. EK94 has a density of 1.73 g/cm^3, R6400 1.75 g/cm^3 and R6510 1.83 g/cm^3. For all three materials, a material analysis using ICP-OES (Induced Coupled Plasma Optical Emission Spectroscopy) was performed. A tendency toward smaller impurities in the newer grade R6510 can be observed with respect to EK94, the oldest grade. However, a material analysis performed with INAA (Instrumental Neutron Activation Analysis) showed much smaller impurities for most trace elements. Due to these discrepancies, the impurity concentrations were set so the calculated activities matched the measured activities.

In addition to the above mentioned material investigations, GDMS (Glow Discharge Mass Spectrometry) was performed to measure the gaseous elements N and Cl. Nitrogen was thought to contribute significantly to the production of ^{14}C via the ^{14}N(n,p)^{14}C reaction. A content of 15 ppm N was measured in EK94. For nitrogen contents larger than 1 ppm, ^{14}C is predominately produced via this reaction, in nuclear reactors [15]. In spite of the missing ^{14}C activity in the calculation as mentioned in the introduction, the effect of adding nitrogen to pure carbon was investigated.

In reactor graphite ^{36}Cl is known ro be produced in large quantities from chlorine. As ^{36}Cl is a long-lived gaseous radioisotopes which is able to migrate out of the final repository, special care is taken when disposing reactor graphite. Furthermore, chlorine outgasses more readily from graphite than from metals due to its porous structure. Therefore chlorine was measured in the material composition and later the resulting ^{36}Cl was calculated (s. section 6).

The results of the material analyses for the isotopes of interest are shown in table 1. The

Table 1. Material analysis from ICP-OES and GDMS.

element	content [ppm]
N	15
Na	13 - 77
Al	6 - 25
Cl	1.1
Ti	0.4 - 1.0
Mn	≈ 0.1

results from ICP-OES for the three graphite grades are presented as ranges.

The Targets E79 and E83 were irradiated for only a few months and disposed as radioactive waste some years later. Samples were taken at disposal time, and the activities were measured shortly after. The measurement of the activity was performed again after 7 and 10 years when short-lived radioisotopes like ^{7}Be could not be detected anymore. More precise values of the sample irradiation and cooling period from both targets can be found in tables 2 and 3.

Table 2. Irradiation history of E79.

period [years]	charge [Ah]
0.6	6.6
9.9	measurement

Table 3. Irradiation history of E83.

period [years]	charge [Ah]
0.2	1.8
7.3	measurement

Target E92 was irradiated to a larger proton fluence of 28.8 Ah. It was irradiated in 4 periods with proton fluences ranging between 6 and 8 Ah with cooling periods of a few months in between. The measurement of the activities was performed 4 years after the last irradiation. In this case, ^{7}Be could still be detected. The irradiation history of E92 can be found in table 4.

Table 4. Irradiation history of E92.

period [years]	charge [Ah]
0.5	6.6
0.3	0
0.5	6.3
0.5	0
0.7	7.4
0.3	0
0.7	8.5
3.8	measurement

3. Experimental determination of specific activities in graphite

In the graphite samples the following radioisotopes were detected: ^{3}H, ^{7}Be, ^{10}Be, ^{14}C, ^{22}Na, ^{26}Al, ^{44}Ti, ^{54}Mn. The gamma emitters ^{7}Be, ^{22}Na, ^{26}Al, ^{44}Ti, ^{54}Mn were detected by a High Purity Germanium detector (HPGe) from ORTEC and analyzed by the software GENIE2000 of Canberra. ^{7}Be could only be detected in the sample of E92 as the cooling times for the other two targets were too long (s. tables 2, 3, 4).

^{10}Be, ^{3}H and ^{14}C were extracted by chemical treatment (wet digestion) as described in reference [16]. A small amount of graphite (20 mg) was boiled in a mixture of acids $H_2SO_4/HNO_3/HClO_4$ and washed out with nitrogen. In the flaks filled with HCL tritium remains in the dissolving solution. In the second flask the carbon dioxide absorbs in the NaOH. ^{3}H and ^{14}C measurements were carried out using liquid scintillation counting (LSC). To determine the absolute activity the yield during the chemical treatment have to be known. Therefore, inactive graphite spiked with an known amount of isotope standard underwent the same chemical treatment as the samples. Afterwards the products were measured under identical conditions. The chemical yields obtained were about 95 % for ^{14}C and about 85 % for ^{3}H [6]. The uncertainty of the ^{14}C measurement is 15 % and the detection limit is 500 Bq/g.

The ^{10}Be content was measured by Multi-Collector Inductively-Coupled- Plamsa Mass Spectrometry (MC-ICP-MS).

4. Calculation

The calculation of the nuclide inventory was done with the Monte Carlo particle transport code MCNPX2.7.0 [7] using the rnucs patch [17] to account for spallation-induced residuals. In this chapter, the method as well as the applied reaction models will be described shortly.

MCNPX2.7.0 can transport a large variety of particles: protons, neutrons, pions as well as ions like deuterons, tritons up to heavy nuclei. As we will see in the following, light ions are important to include as they can make significant contributions to the activation. For the nuclear reactions of all particles, physics models have to be used when no cross section tables are available or selected. One class of nuclear reaction models are based on the intranuclear cascade (INC) mechanism followed by evaporation or fission. These models are usually applicable from about 150 MeV to a few GeV. However, often they are used for lower energies from 20 MeV on, where some cross section tables stop working. In INC, the energy of the primary particle is dissipated in the nucleus to several nucleons by nucleon-nucleon collisions until equilibrium is reached. In the beginning of this phase, the so called pre-equilibrium, particles, which receive a large energy transfer by direct interaction, can leave the nucleus with large kinetic energy in forward direction (relative to the incoming particle). Later, the nucleus is in an excited state with an often nonzero angular momentum. Both energy and angular momentum can be released by either emission of light particles like photons, neutrons, deuterons etc. (evaporation) or by splitting the nucleus in two (fission). Large particles like Mn can be emitted during the evaporation process as well, but with lower probability. As the particles are emitted at relatively low energy in the evaporation phase, they are isotropically distributed. Fission occurs mainly for heavier nuclei like Pb as these are more affected by surface vibrations. The evaporation and fission is often treated by separate models, to which the INC code is coupled. The INC code BERTINI [8, 9] is often coupled to the evaporation code DRESNER [10] and the fission model RAL [11]. The DRESNER code is based on Weisskopfs statistical model. This combination has been the default choice in MCNPX for neutrons and protons of energy less than 3.5 GeV for years. As BERTINI can handle only protons and neutrons, reactions containing deuterons, tritons and α-particles up to 1 GeV are handed over to ISABEL [18]. ISABEL can also replace BERTINI completely.

MCNPX2.7.0 contains an outdated version of INCL/ABLA (INCL4.2/ABLAV3p), which has a couple of well-known shortcomings. Therefore, in the modified version that was used, INCL was upgraded to INCL4.6 [12] and ABLA to ABLA07 [13]. Up to now this model is not yet in one of the distributed versions of MCNPX or MCNP. With the help of J.-Chr. Davide, INCL4.6/ABLA07 was implemented in the version of MCNPX2.7.0 in such a way that it could be run on multiple processors. As the name implies, the INCL models are also based on INC. The new INCL4.6 model handles not only neutrons, protons, pions as in INCL4.2, but also deuterons, tritons, ^{3}He, α-particles, and light ions up to mass number A = 8. These particles cannot only induce nuclear reactions, but can also be emitted in all phases of the physics model. The light ions are formed by coalescence in phase space from neutrons and protons. ABLA07 itself is able to generate neutrons, protons, deuterons, tritons, ^{3}He, α-particles, and intermediate mass fragments (IMF) by break-up, fission or evaporation. In addition, when using INCL4.6/ABLA07, metastable isotopes can now be handled in MCNPX2.7.0 according to the PHTLIB file as it is done for the default cross section choice.

In the following simulations, the two above mentioned models are used for all particle interactions except for neutrons with energies less than 20 MeV. Below 20 MeV, tabulated cross sections from the ENDF/B-VII [19] dataset or similar versions are applied. For later evaluation of the nuclide inventory, the neutron flux spectra up to 20 MeV are written to the output file of MCNPX for the cells of interest. For the other interactions controlled by the physics models, a lot of information about each reaction is usually written into a binary file called "histp". This file can become several GB in size. Since the calculation of the nuclide inventory requires only the production rates, this information is directly stored in the output file of MCNPX and the histp-file can be dismissed. This information is made available in the output file by using the rnucs-card [17]. With the help of a PERL script (activation script [20]), all information including the irradiation history and cooling times is passed to build-up and decay codes like CINDER1.05 [21] or FISPACT10 [22] which can use the cross section library EAF-2010 [22]. In the following evaluation, the results using FISPACT10/EAF-2010 are presented. However, a check with CINDER1.05, which comes with its own library, did not lead to significant different results.

Most simulation runs were performed for the sample from Target E92: 11 runs using BERTINI-DRESNER-RAL and 36 runs using INCL4.6/ABLA07. The material composition and the transported particles were changed between the runs (see below). Up to 10^8 primary protons were needed to ensure a statistically significant result. When reactions with impurities were not being sampled sufficiently due to their low concentrations, the option "impure" of the rnucs-card became important. This ensures that all isotopes of the material composition undergo reactions with the same probability. Concentration is taken into account by applying the proper weight to the tally score to ensure a fair game is played and the result means are not biased.

The calculations were performed on a medium-performance cluster at PSI. The number of cores used was between 24 and 72. The core-hours per primary particle depend strongly on the number of particles transported and the number of isotopes in the material composition. On 24 cores, MCNPX2.7.0 using INCL4.6/ABLA07 with the extended list of particles and pure carbon needs 6 h for 10^8 primary protons. BERTINI-DRESNER-RAL, known as the faster physics model, was in fact a factor 2 slower as this compilation of MCNPX2.7.0 was done without compiler optimizations.

4.1. Geometrical model

For the activation calculations, a precise geometrical model were used (see figure 2). The reason for doing this is twofold. First, this model will be used also for other purposes like calculating the energy deposition on the cooled water plates as well as comparing measured and calculated activity of a concrete sample located 2 m above beam height. Second, during the former investigations regarding the nuclide inventory [5], it was thought that the large discrepancy between calculation and measurement regarding the ^{14}C activity is caused by backscattered neutrons, which were in the early analysis not taken into account as the geometrical model consists just of a graphite block. The backscattered neutrons would have low energy where the neutron capture cross section on ^{13}C producing ^{14}C is large. In figure 2, copper parts are shown in blue. They are almost all cooled by water, which is not taken into account in the simulation. Steel parts are shown in green. The shielding around the target chamber are made from normal carbon steel.

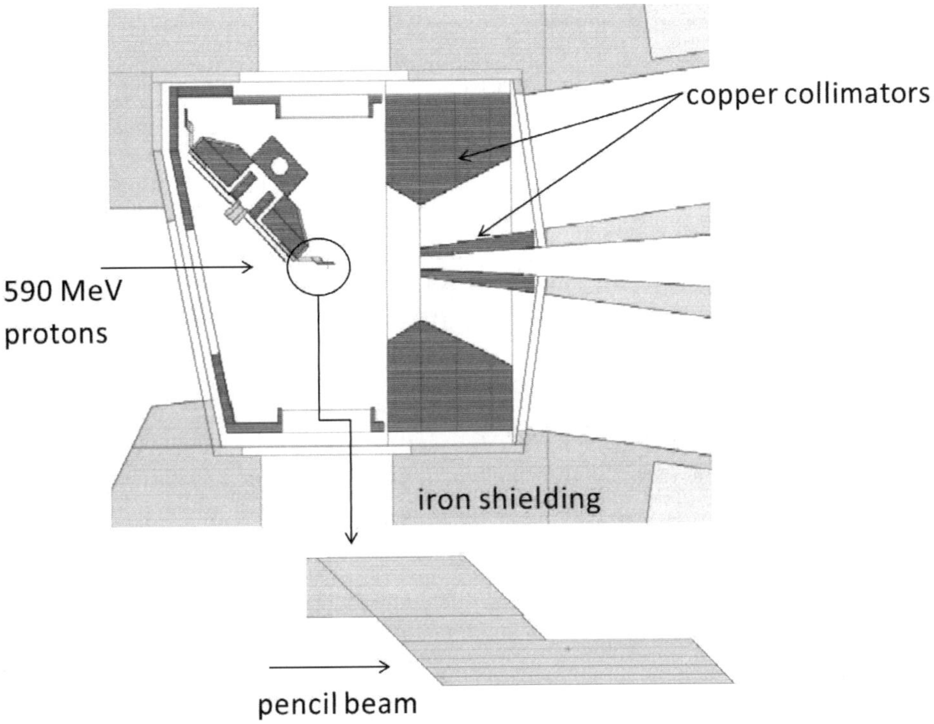

Figure 2. Geometry in MCNPX for the current calculations.

The rim of the graphite wheel sliced into 5 pieces in horizontal direction. For both the 40 mm and 60 mm target, the rim is 6 mm wide. The beam is centered on the rim and usually has a Gaussian distribution with standard deviations between 0.5 and 1 mm. In the vertical plane, the beam is a bit wider than in the horizontal direction. Since the beam parameters and the position on the target vary, the simulation assumes a pencil beam at the rim center for simplicity.

5. Results

In this section, the calculations and results obtained under different settings will be presented and compared to the experimental values. Except for ^{10}Be, the experimental results were already published in [6]. Most investigations were done on E92 since this is the target with the highest proton fluence and the shortest cooling time. A summary of all results will be given at the end of this report.

5.1. ^{14}C activity

The main investigations were done concerning the ^{14}C activity due to the large discrepancy of 2 orders of magnitude seen in the past. First, using BERTINI-DRESNER-RAL in MCNPX2.7.0 this result could be confirmed. The ratio between the calculated and measured activities on E92 is 1/71.5. The uncertainty for the measured ^{14}C activity is 15 %. By default, neutrons, protons, charged and neutral pions, deuterons, tritons, ^{3}He, and α-particles were taken into account. For this and the following results a pure carbon target is assumed except when stated otherwise. The next trial was to add 15 ppm nitrogen to the pure natural carbon since this concentration was measured in EK94. This increased the ^{14}C activity by 35 % - not enough to explain the large discrepancy.

Using INCL4.6/ABLA07 on a pure carbon target, the ratio of the activities from Monte Carlo and measurement increased to 0.92, which is an enormous improvement compared to BERTINI-DRESNER-RAL, and even within the error bars of the measurement. To figure out the reaction mechanism for the production of ^{14}C, a simulation was performed with the "standard" particles neutrons (n), protons (p) and pions (π) for both models on pure carbon. This lead to very similar results for both models: using BERTINI-DRESNER-RAL 58 Bq/g and with INCL4.6/ABLA07 80 Bq/g for ^{14}C were obtained. The value for BERTINI-DRESNER-RAL is very similar to the one obtained with the more extended list of particles (65 Bq/g) whereas the activity calculated with INCL4.6/ABLA07 and the "standard" particles is 53 times smaller. This means that the main production mechanism of ^{14}C are neither the direct production via high energetic protons nor neutron caption on ^{13}C. Therefore, one or more of the remaining particles, deuterons (d), tritons (t), ^{3}He, α-particles have to be responsible for the sizeable increase of the ^{14}C production. Next, simulations with INCL4.6/ABLA07 were performed including the "standard" set of particles plus *one* particle from the remaining list. The results are shown in table 5.

Table 5. Activity of ^{14}C using INCL4.6/ABLA07 and the given list of particles transported.

particles	^{14}C activity [Bq/g]
n,p,π	80
+ t	2529
+ d	1100
+ α	592
+ ^{3}He	122

From these results, clearly tritons and deuterons produce most of the ^{14}C. Therefore, the triton and deuteron cross sections and flux spectra using BERTINI-DRESNER-RAL as well as INCL4.6/ABLA07 were extracted from MCNPX2.7.0. The comparison of the cross sections on ^{12}C as well as ^{13}C are shown in figure 3.

In figure 3, the cross sections extracted from BERTINI (dashed line) are always lower than the ones representing INCL4.6 (solid line). Large differences of about three orders of magnitude can be found for tritons on ^{12}C and deuterons on ^{13}C. Although the cross sections of tritons on ^{13}C is larger than ^{12}C, the abundance of ^{13}C is only 1.1 %. The production of ^{14}C by deuterons on ^{12}C is negligible in both models and is not shown in figure 3.

For the production of ^{14}C, not only the cross sections are important but also the number of incident particles and their energy distribution. The flux spectra of tritons and deuterons as a function of energy are shown in figure 4. The results for BERTINI-DRESNER-RAL (dotted) look similar in shape for both species as well as the spectra using INCL4.6/ABLA07 (solid line). The spectra from INCL4.6/ABLA07 have an additional tail at higher energy, which becomes

IOP Conf. Series: Journal of Physics: Conf. Series **1046** (2018) 012003 doi:10.1088/1742-6596/1046/1/012003

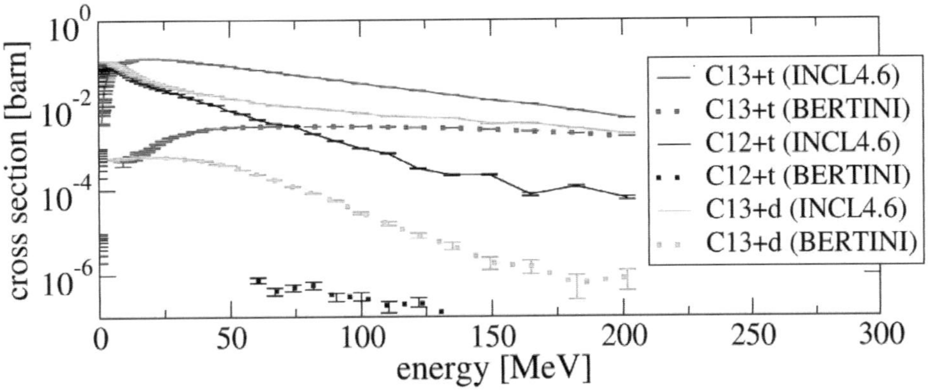

Figure 3. Triton and deuteron cross sections for the production of ^{14}C on ^{12}C and ^{13}C using BERTINI (dashed) and INCL4.6 (solid).

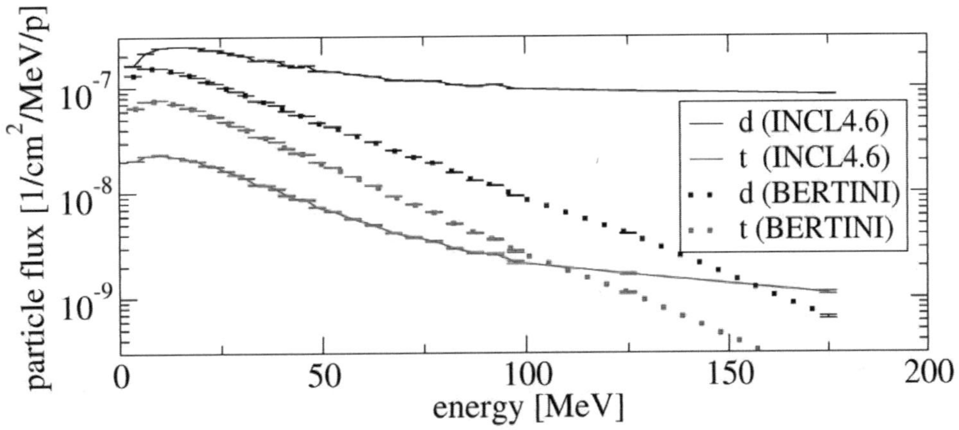

Figure 4. Triton and deuteron flux spectra using BERTINI-DRESNER-RAL and INCL4.6/ABLA07.

clearly visible for energies larger than 100 MeV. This is due to the emission of higher energetic particles during the pre-equilibrium phase. Interestingly, the triton flux from INCL4.6/ABLA07 is significantly smaller for energies less than 100 MeV than the corresponding spectrum from BERTINI-DRESNER-RAL.

From the triton and deuteron cross sections and energy distributions, significant differences in the production of ^{14}C are obvious, which are responsible for the increased ^{14}C activity using INCL4.6/ABLA07 and the good agreement to the experimental data. For E79 and E83, the ratio of the ^{14}C activity from INCL4.6/ABLA07 to the measurement are 0.50 and 1.16, respectively. The ^{14}C activity in E83 was already a factor 2 below the detection limit, however, the measurement is in good agreement with the prediction. Averaging the results from the three samples leads to 0.86 for pure carbon. The largest effect would be a 15 ppm nitrogen contribution which would result in 30 % more ^{14}C and increase the ratio to 1.1. However, it is not clear, if the nitrogen content is the same in all targets as the nitrogen content was measured in EK94 only. Particularly, R6510 has a 7 % higher density and therefore less porosity, which leads to the expectation of smaller amounts of nitrogen. In addition one has to consider that nitrogen might be completely outgassed during operation when the graphite is heated to about 1500 ^{o}C.

5.2. Be activity

In all three target samples, ^{10}Be was determined experimentally. The short-lived ^{7}Be isotope could be measured in E92 only. For E92, simulations using pure carbon were performed with both physics models. BERTINI-DRESNER-RAL agrees well with the experimental data for both Be-isotopes. The ratio Monte Carlo to measurement is 1.32 and 1.01 for ^{7}Be and ^{10}Be, respectively. The MCNPX2.7.0 simulation with INCL4.6/ABLA07 leads to higher ratios: 1.84 and 1.94 for ^{7}Be and ^{10}Be. In case of ^{7}Be, INCL4.6/ABLA07 results in 40 % more activity copmared to BERTINI-DRESNER-RAL. For ^{10}Be the increase is even higher at 90 %.

A similar trend is observed for the other two targets. Here only INCL4.6/ABLA07 simulations were performed as the result for the calculation with BERTINI-DRESNER-RAL can be estimated from the comparison to INCL4.6/ABLA07 done for E92. For E79, INCL4.6/ABLA07 leads to 1.48 for the ^{10}Be activity ratio, whereas for E83 the ratio is 3.02. The larger discrepancy might have also to do with the fact that E83 was irradiated to the lowest charge, and therefore the uncertainty in the measurement might be larger. For BERTINI-DRESNER-RAL, a ratio of about 1.6 can be expected, which would be an acceptable agreement.

5.3. Tritium activity

^{3}H activity was measured in all three targets. In addition, the graphite samples were sliced horizontally with regard to the beam direction, and the tritium was measured in each slice. In the 6 mm wide rim of the target, 3 to 5 slices could be made, each of which have different widths. In the simulation, 5 slices symmetrically around the center (s. figure 2) with a width of 6/5 mm were modeled. In the simulation, a pencil beam is assumed whereas the real beam has a Gaussian distribution in vertical and horizontal direction with standard deviation of 0.5 to 1 mm (FWHM = 1.2 to 2.3 mm). Therefore, it is expected that the calculated distribution is smaller and particularly the peak maximum higher than the measured distribution.

In figure 5, the experimental results for the samples from E79 are compared to the MCNPX2.7.0 simulation using INCL4.6/ABLA07. Two samples were cut into 3 and 4 slices. The two measurements agree well to each other. The width of each slice is indicated by the horizontal error bar. Some slices are smaller, others larger. The results are not corrected for bin sizes as then a shape of the distribution would need to be known or assumed. As expected, the experimental distribution is wider than the calculated one assuming a pencil beam. The experimental distributions seem to be slightly off center. The ratio of the Monte Carlo result and the measurement is 1.11, i.e. both are in very good agreement.

In figure 6, the two measured tritium distributions in E83 are compared to the simulation. The experimental samples are cut in three and five slices and agree well to each other. The maximum is in the center whereas a bump is seen on the left. This might be due to an asymmetric beam distribution, or due to the superposition of different positions of the beam. The second case is more likely as the beam is indeed shifted to one or the other side depending which beam lines (left or right from Target E) has priority. The activity ratio for tritium from simulation to measurement is 0.79. Again, a reasonable agreement is achieved.

In figure 7, the results for the samples from E92 are shown. Four samples were cut into three to five slices. However, none of the experimental distributions shows a clear peak. In addition, the measured tritium activity is much smaller than predicted from the calculation. The ratio of simulated and measured activity is 1/12, i.e. 90 % of the tritium is missing. Using BERTINI-DRESNER-RAL instead of INCL4.6/ABLA07 would lead to 40 times less tritium. In general, due to the high operation temperature ($\approx$ 1500 ^{o}C) of the target, it is possible, that tritium

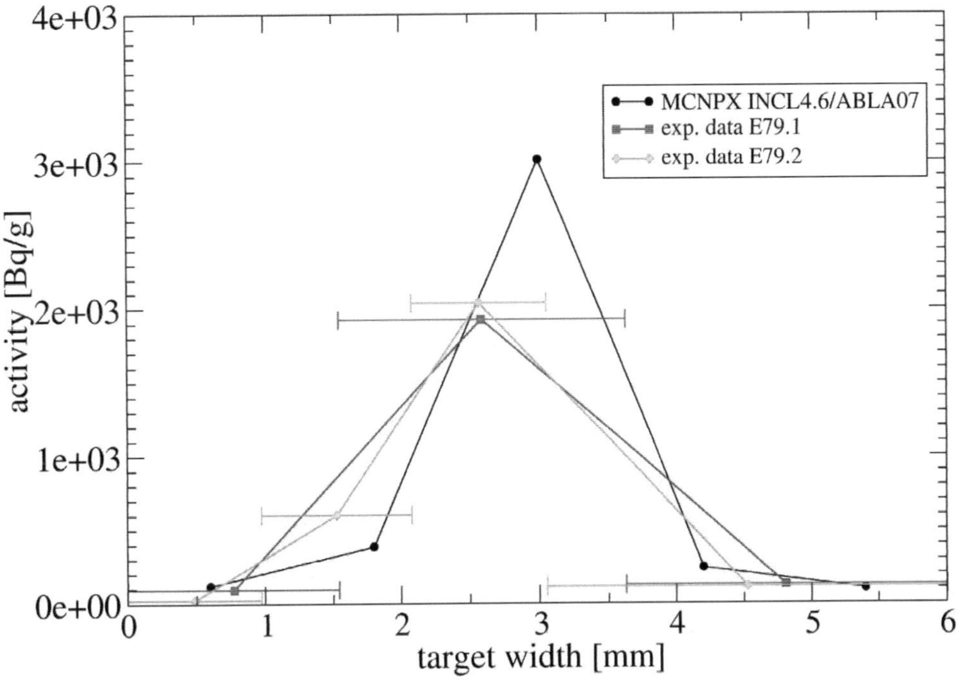

Figure 5. Comparison of the simulated (using INCL4.6/ABLA07) and measured tritium activity as a function of the target width for E79.

already outgassed. However, the question is, why this happens in E92 only and not in E79 and E83. E92 was operated at a bit higher current. However, since the target is cooled by radiation, the operation temperature is changed only slightly (about 50 to 100 ^oC). However, in metals hydrogen starts to diffuse out at temperatures larger than 250 ^oC. Another difference is that E92 is made from a different grade than E79 and E83. Although the E92 graphite grade R6510 is more radiation hard due to smaller grains, the target E92 suffered from the largest beam charge, 4 times more than E79 and 16 times more than E83. It is known that particle radiation changes the structure of the material. The charge of about 29 Ah corresponds to 3 DPA (Displacements Per Atom). This means that every atom in the graphite structure is moved three times. Many defects in the structure occur, which facilitates the movement and outgassing of hydrogen.

5.4. Activities from impurities

As already mentioned, the activities of the gamma emitters ^{22}Na, ^{26}Al, ^{44}Ti and ^{54}Mn were also detected. These radioisotopes are not produced by carbon, but from impurities with larger proton number like Na, Al, Ti and Mn. These elements were indeed detected in the material. However, first they vary with graphite grade, and second they lead to much larger activities than the carbon-born isotopes. Therefore, it was decided to apply another analysis method. For the mentioned elements, simulations were performed with pure carbon plus 10 ppm of one element as impurity. The resulting activity of the leading radioisotope (e.g. ^{22}Na when using Na as impurity) was then compared to the measured activity. From this, the content of the impurity of the sample was deduced and compared to the abundance in the material analysis measured with ICP-OES. The investigations regarding the activities from impurities were performed for the samples from E92 and with the model INCL4.6/ABLA07 only. In table 6 the second column

IOP Conf. Series: Journal of Physics: Conf. Series **1046** (2018) 012003 doi:10.1088/1742-6596/1046/1/012003

contains the experimental data from E92 whereas the last column shows the results from the simulation using besides carbon 10 ppm of the element indicated in the third column.

Table 6. Experimental and simulated activities for the radioisotopes from impurities as indicated.

nuclide	exp. activity [Bq/g]	material comp.	MC activity [Bq/g]
^{22}Na	52	C + 10 ppm Na	1.52 10^5
^{26}Al	0.3	C + 10 ppm Al	1.87
^{44}Ti	16	C + 10 ppm Ti	1454
^{54}Mn	2.5	C + 10 ppm Mn	2.14 10^4

As can be already seen from table 6, a small amount of Na or Mn leads to a large activity of ^{22}Na and ^{54}Mn, respectively . In addition, 10 ppm Al also produces 4.1 10^4 Bq/g ^{22}Na and 10 ppm Mn leads to 957 Bq/g ^{44}Ti. Therefore, the amount of the impurities estimated from the comparison of measured and simulated activities in table 6 have to be seen as maximum values. These values are summarized in table 7 and are compared to the ICP-OES material analysis of R6510.

Further, other impurities also lead to sometimes significant contributions to the radioisotopes listed in table 7. For this a simulation was performed adding 10 ppm Fe to pure Carbon. As expected, a significant amount of ^{54}Mn (9500 Bq/g) and ^{44}Ti (1300 Bq/g) are produced.

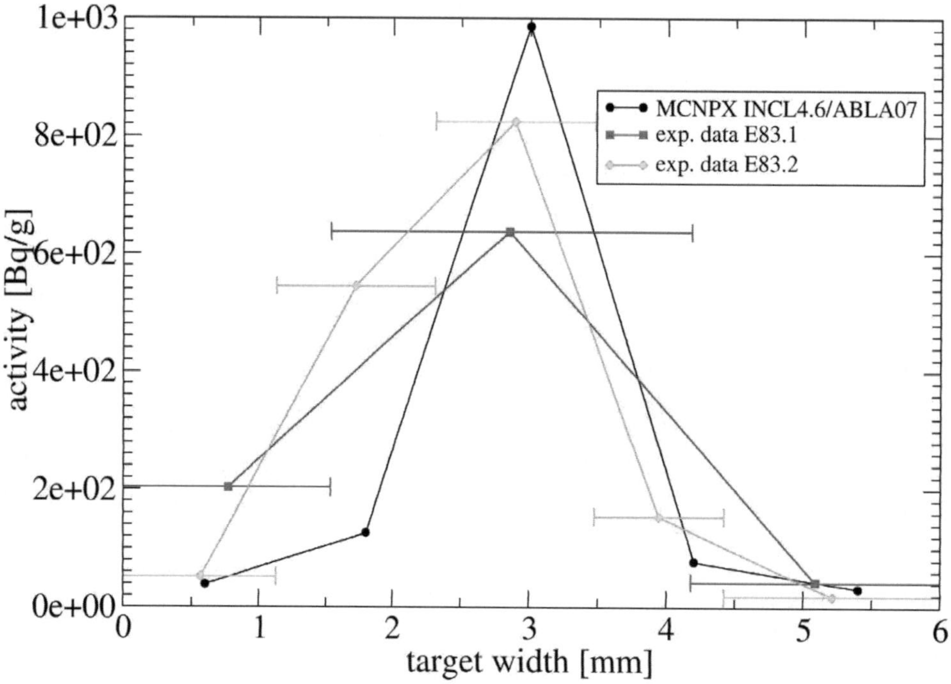

Figure 6. Comparison of the simulated (using INCL4.6/ABLA07) and measured tritium activity as a function of the target width for E83.

IOP Conf. Series: Journal of Physics: Conf. Series **1046** (2018) 012003 doi:10.1088/1742-6596/1046/1/012003

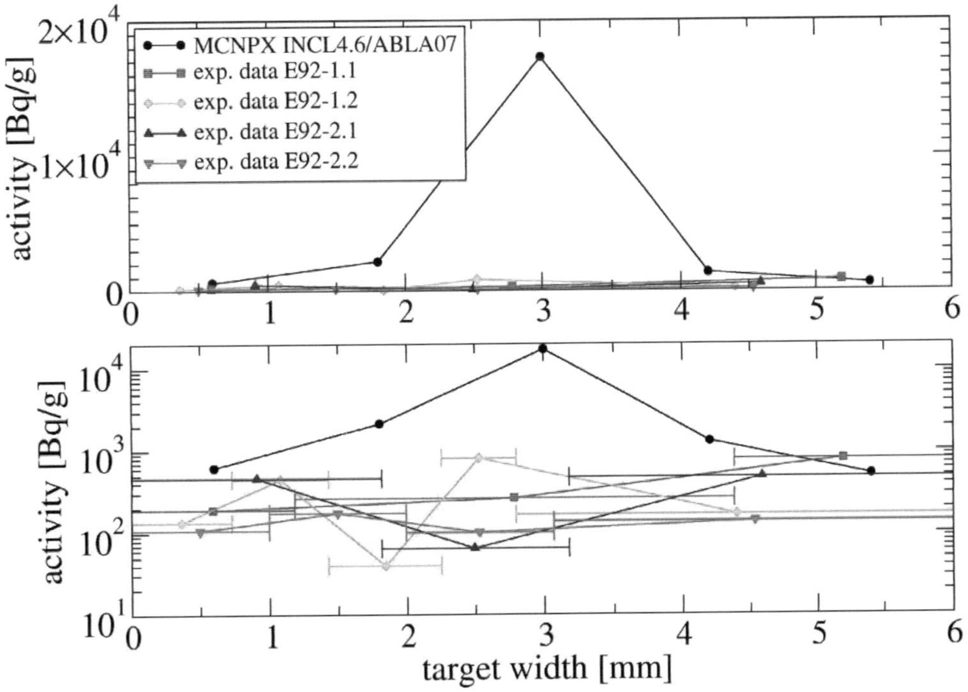

Figure 7. Comparison of the simulated (using INCL4.6/ABLA07) and measured tritium activity as a function of the target width for E92.

Comparing to table 6, this is half of the ^{54}Mn activity obtained with 10 ppm Mn and almost the same amount of ^{44}Ti as produced with 10 ppm Ti.

Table 7. Content of trace elements in the first column deduced from the comparison of calculational and experimental activity of the radionuclides in the second column as well as compared to the ICP-OES measurement.

element	nuclide	(MC/exp.) [ppm]	ICP-OES [ppm]	ICP-OES/MC
Na	^{22}Na	0.0034	13	3805
Al	^{26}Al	1.61	5.8	4
Ti	^{44}Ti	0.11	0.66	6
Mn	^{54}Mn	0.0012	0.12	103

The third column of table 7 was calculated by comparing the measured and simulated activities in table 6. In the Monte Carlo simulation, 10 ppm of the trace element was always used. From the measured ^{22}Na activity just 0.0034 ppm should be present in the sample while 13 ppm Na were measured using ICP-OES. However, Na is present everywhere as an impurity and therefore it is likely that the large amount measured in the sample is a contamination during the measurement. The measured amount of 0.12 ppm Mn is already quite small. However, the concentration deduced with the help of the simulation is even a factor 100 smaller. This very small amount would not be measurable by ICP-OES. In general, determining impurities in the sub-ppm region is very difficult and requires another method like ICP-MS (Induced Coupled Plasma Mass Spectrometry) as well as very clean conditions. Therefore, it can be concluded that the graphite used for target E is much purer than assumed from the material analysis. This also explains the observation that after a few years of cooling the calculated dose rates of all former graphite targets were much higher than measured. When the ^{7}Be is decayed, the dose rate after a few years is mainly determined by ^{22}Na, which is according to table 7 about 3 orders too high in the formerly assumed material composition.

6. Summary results and conclusions

The results for the comparison of measured and calculated activities of the graphite target E are summarized in the tables 8, 9, 10 and 11.

Table 8. Summary of the results for measured and calculated activities on samples from E79. The simulation was performed with MCNPX2.7.0 using INCL4.6/ABLA07.

nuclide	exp. activity [Bq/g]	MC activity [Bq/g]	MC/exp.
^{3}H	6.94E+08	7.73E+08	1.11
^{10}Be	4004	5943	1.48
^{14}C	2041	1022	0.50

Table 9. Summary of the results for measured and calculated activities on samples from E83. The simulation was performed with MCNPX2.7.0 using INCL4.6/ABLA07.

nuclide	exp. activity [Bq/g]	MC activity [Bq/g]	MC/exp.
^{3}H	3.19E+08	2.52E+08	0.79
^{10}Be	546	1650	3.02
^{14}C	244	284	1.16

Table 10. Summary of the results for measured and calculated activities on samples from E92. The simulation was performed with MCNPX2.7.0 using INCL4.6/ABLA07.

nuclide	exp. activity [Bq/g]	MC activity [Bq/g]	MC/exp.
^{3}H	3.60E+08	4.43E+09	12.30
^{7}Be	594	1092	1.84
^{10}Be	13567	26300	1.94
^{14}C	4648	4270	0.92

The uncertainty for the ^{14}C measurement is 15 % and its detection limit is 500 Bq/g, meaning the ^{14}C activity of E83 measurement is below this limit. However, all measurements at the three targets regarding ^{14}C are for the first time well reproduced by using the cross section model INCL4.6/ABLA07 in MCNPX2.7.0. The default model, BERTINI-DRESNER-RAL used in

Table 11. Summary of the results for measured and calculated activities on samples from E92. The simulation was performed with MCNPX2.7.0 using BERTINI-DRESNER-RAL.

nuclide	exp. activity [Bq/g]	MC activity [Bq/g]	MC/exp.
^{3}H	3.60E+08	1.45E+10	40.17
^{7}Be	594	784	1.32
^{10}Be	13567	13680	1.01
^{14}C	4648	64.92	0.01

former analyses underpredicts the ^{14}C activity by two orders of magnitude. The main reaction responsible for the large ^{14}C production is neither the neutron capture in ^{13}C nor a contamination with nitrogen. Reactions with tritons and deuterons on ^{12}C lead to the additional ^{14}C activity. This indirect mechanism required 10^8 primary protons to be simulated in order to achieve good result statistics.

From the results of E92, it seems that ^{3}H outgassed during operation of the target. This is supported by the flat distribution of ^{3}H along the target width (horizontal direction), whereas in E79 and E83, the peak of the tritium activity is clearly visible around the center of the target width. Candidates for an explanation are the different graphite grade or radiation damage.

Comparing the activities from the simulations using BERTINI-DRESNER-RAL and INCL4.6/ABLA07 for the Be-isotopes, ^{7}Be and ^{10}Be, the default model of MCNPX2.7.0 fits the experimental data better. INCL4.6/ABAL07 is higher than the measurement for all cases, and about 40 % higher for ^{7}Be activity and 90 % for ^{10}Be activity than the simulation using BERTINI-DRESNER-RAL.

The analysis of the activities from impurities revealed a much purer graphite than assumed from the ICP-OES material analysis. The impurities seem to be in the sub-ppm region, which might require another measurement method and procedures to avoid contamination. Before the investigation in this report, it was not expected that impurities on the sub-ppm level would lead to a measurable activity. Not measured, but discussed in the introduction, is the activity of ^{36}Cl, which is thought to be a problematic nuclide due to its long life time and its ability to migrate out of the component. In the EK94 grade graphite, 1.1 ppm Cl was determined. Using MCNPX2.7.0 and INCL4.6/ABLA07 this would lead to 0.12 Bq/g ^{36}Cl. The activity is smaller than expected due to the fact that ^{36}Cl is produced by capturing thermal neutrons. This is due to the fact that few thermal neutrons are produced in the meson target compared to a nuclear reactor.

Acknowledgement
We would like to thank Tanja Wieseler und Sabrina Lüthi for the chemical preparation of the graphite samples and the measurements of the activities. Further, we are grateful to Dr. Jean-Christophe Davide for giving us access to INCL4.6/ABLA07 and helping us with the implementation into MCNPX2.7.0.

References

[1] Argentini M and Weinreich R 2000 *Measurement of radionuclide contents in activated graphite* (Switzerland: PSI) p 159 PSI Scientific and Technical Report, Volume I: Particles and Matter

[2] Argentini M and Weinreich R 2002 *Determination of Tritium and Carbon-14 in accelerator waste* (Jülich, Germany: R. Odoj, J. Baier, P. Brennecke, K. Kühn) pp 393 – 396 4th International Seminar on Radioactive Waste Products RADWAP

[3] Prael R and Lichtenstein H (eds) 1989 *User Guide to LCS: The LAHET Code System* LA-UR-89-3014 (Los Alamos, USA: LANL)
[4] Pelowitz D (ed) 2005 *MCNPX user's manual - version 2.5.0* LA-CP-05-0369 (Los Alamos, USA: LANL)
[5] Kiselev D 2006 unpublished
[6] Schumann D, Lüthi S, Stowasser T, Kiselev D and Teichmann S 2012 *14C and 3H determination in graphite wheels* (Tucson, USA: University of Arizona) p 370 Radiocarbon
[7] Pelowitz D (ed) 2011 *MCNPX user's manual - version 2.7.0* LA-CP-11-00438 (Los Alamos, USA: LANL)
[8] Bertini H 1963 *Phys. Rev.* **131** 1801 – 1821
[9] Bertini H 1969 *Phys. Rev.* **188** 1711 – 1730
[10] Dresner L 1981 *EVAP-A Fortran Program for Calculating the Evaporation of Various Particles from Excited Compound Nuclei* ORNL-TM-7882 (USA: ORNL)
[11] Atchison F 1969 *Nucl. Instr. Meth.* B **259** 909 – 932
[12] Boudard A, Cugnon J, David J C, Leray S and Mancusi D 2013 *Phys. Rev.* C **87** 014606 – 28
[13] Kelic A, Ricciardi M and Schmidt K H 2008 *ABLA07 - Towards a Complete Description of the Decay Channels of a Nuclear System from Spontaneous Fission to Multifragmentation* (IAEA) pp 181 – 222 (*IAEA report INDC(NDS)*- no 0530)
[14] Baumgarten C, Barchetti A, Einenkel H, Goetz D and Schmelzbach P 2011 *Rev. Sci. Instrum* **82** 053304
[15] Heikola T 2014 *Leaching of 14C in repository conditions* (*VTT Technology* no 157) (Finland: VTT Technical Research Centre of Finland)
[16] Schumann D, Neuhausen J, Horn S, Kubik P and Günther-Leopold I 2008 *Radiochim Acta* **96** 31 – 34
[17] Gallmeier F and Wohlmuther M 2015 *The Residual Nuclei Patch of MCNPX2.6.0 and MCNPX2.7.0* TM-85-15-03 (Switzerland: PSI)
[18] Yariv Y and Fraenkel Z 1981 *Phys. Rev.* C **24** 488 – 494
[19] Chadwick M 2006 *Nuclear Data Sheets* **107** 2931 – 3060
[20] Gallmeier F and Wohlmuther M 2008 *ACTIVATION SCRIPT VERSION 1.0 USER GUIDE* ORNL-TM-031 (USA: ORNL)
[21] Wilson W, England T, George D, Mur W and Young P 1995 *Recent Developments of the CINDER'90 Transmutation Code and Data Library for Actinide Transmutation Study* (Versailles, France: IAEA) pp 848 – 854 Proc. GLOBAL'95 Int. Conf. on Evaluation of Emerging Nuclear Fuel Cycle Systems, INIS 29 (49)
[22] Sublet J C, Packer L, Kopecky J, Forrest R, Koning A and Rochman D 2010 *The European Activation File: EAF-2010 neutron induced cross section library* EASY Documentation Series CCFE-R (10) 05 (United Kingdom: EURATOM/CCFE Fusion Association)

Measurements and FLUKA Simulations of Bismuth, Aluminium and Indium Activation at the upgraded CERN Shielding Benchmark Facility (CSBF)

E. Iliopoulou[1,2], P. Bamidis[2], M. Brugger[1], R. Froeschl[1], A. Infantino[1], T. Kajimoto[3], N. Nakao[4], S. Roesler[1], T. Sanami[5], A. Siountas[2], H. Yashima[6]

[1] CERN, 1211 Geneva 23, Switzerland
[2] Medical Physics Laboratory, School of Medicine, Aristotle University of Thessaloniki, 54124 Thessaloniki, Greece
[3] Hiroshima University, 1-4-1 Kagamiyama, Higashi-Hiroshima 739-8527, Japan
[4] Shimizu Corporation, Etchujima 3-4-17, Koto-ku, Tokyo 135-8530, Japan
[5] High Energy Accelerator Research Organization-KEK, Oho, Tsukuba 305-0801, Japan
[6] Research Reactor Institute, Kyoto University, 2-1010 Asashiro-nishi, Kumatori, Sennan, Osaka 590-0494, Japan

E-mail: `elpida.iliopoulou@cern.ch`

Abstract. The CERN High energy AcceleRator Mixed field (CHARM) facility is situated in the CERN Proton Synchrotron (PS) East Experimental Area. The facility receives a pulsed proton beam from the CERN PS with a beam momentum of 24 GeV/c with $5{\cdot}10^{11}$ protons per pulse with a pulse length of 350 ms and with a maximum average beam intensity of $6.7{\cdot}10^{10}$ protons per second. The extracted proton beam impacts on a cylindrical copper target. The shielding of the CHARM facility includes the CERN Shielding Benchmark Facility (CSBF) situated laterally above the target that allows deep shielding penetration benchmark studies of various shielding materials. This facility has been significantly upgraded during the extended technical stop at the beginning of 2016. It consists now of 40 cm of cast iron shielding, a 200 cm long removable sample holder concrete block with 3 inserts for activation samples, a material test location that is used for the measurement of the attenuation length for different shielding materials as well as for sample activation at different thicknesses of the shielding materials. Activation samples of bismuth, aluminium and indium were placed in the CSBF in September 2016 to characterize the upgraded version of the CSBF. Monte Carlo simulations with the FLUKA code have been performed to estimate the specific production yields of bismuth isotopes (^{206}Bi, ^{205}Bi, ^{204}Bi, ^{203}Bi, ^{202}Bi, ^{201}Bi) from ^{209}Bi, ^{24}Na from ^{27}Al and ^{115m}I from ^{115}I for these samples. The production yields estimated by FLUKA Monte Carlo simulations are compared to the production yields obtained from γ-spectroscopy measurements of the samples taking the beam intensity profile into account. The agreement between FLUKA predictions and γ-spectroscopy measurements for the production yields is at a level of a factor of 2.

1. Introduction

The CERN High Energy AcceleRator Mixed field facility (denoted CHARM) has been constructed in the CERN Proton Synchrotron (PS) East Experimental Area in 2014 [1]. The facility receives a pulsed proton beam from the CERN PS with a beam momentum of 24 GeV/c with $5{\cdot}10^{11}$ protons per pulse with a pulse length of 350 ms and with a maximum average beam intensity of $6.7{\cdot}10^{10}$ p/s.

The extracted proton beam from the PS impacts on a cylindrical Copper or Aluminium target and the created secondary radiation field is used to test electronics equipment installed at predefined test positions.

The shielding of the CHARM facility [2] also includes the CERN Shielding Benchmark Facility (CSBF) situated laterally above the target [3]. This facility allows deep-penetration benchmark studies of various shielding materials [4, 5, 6, 7]. The CHARM facility at beam line level is illustrated in figure 1 indicating the direction of the beam coming from the Proton Synchrotron (PS), the CHARM target, the target alcove for storing the target during access and movable shielding walls.

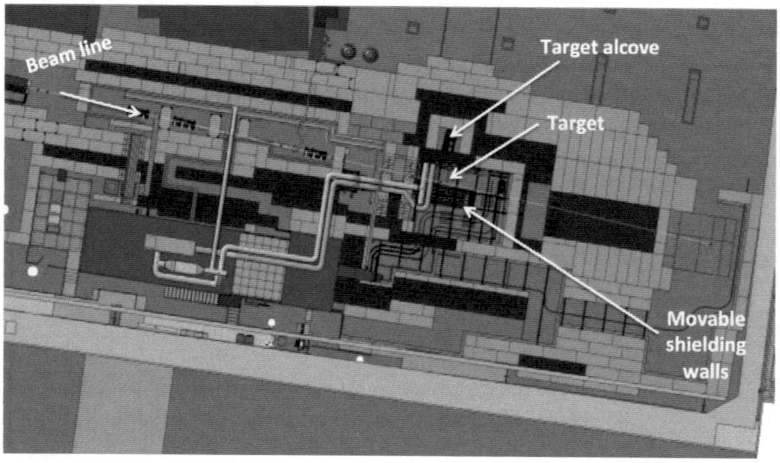

Figure 1. Horizontal integration drawing of the CHARM facility at beam line level. The CSBF is located laterally above the CHARM target.

An activation experiment campaign was performed in 2015 at the CSBF for the characterization of the facility [8, 9, 10] and, based on our experience, we decided to upgrade the CSBF in 2016 in order to simplify the exploitation of the facility and to integrate new functionalities. In order to understand and characterize the radiation fields in the upgraded CSBF, activation samples were placed in the CSBF in September 2016. Monte Carlo simulations with the FLUKA code [11, 12] have been performed to estimate the specific production yields of several bismuth isotopes, ^{24}Na and ^{115m}I for these samples. This paper describes the comparison between the estimated values from FLUKA and the activation measurements performed in September 2016 with bismuth, aluminium and indium disk samples of different sizes in the upgraded CSBF.

2. Design of the CERN Shielding Benchmark Facility (CSBF)

The CERN Shielding Benchmark Facility (CSBF) has been significantly upgraded during the extended technical stop at the beginning of 2016. The CSBF upgrade allows for easier manipulation and for having more exploitation possibilities of the facility [10, 13]. The design of the upgraded CSBF was based on FLUKA simulations.

During the operational period of 2016, the CSBF consisted of 40 cm cast iron shielding, 360 cm of standard concrete, barite concrete and cast iron shielding that are part of the three main possible configurations of the CSBF. These three main possible configurations allow measurement at the removable sample holder concrete block (which is also the nominal configuration of the facility during the nominal CHARM facility operation), on the CSBF platform and in the shielding material test location.

2.1. Removable sample holder concrete block

The removable sample holder concrete block was needed for the facilitation of the handling procedure of the activation samples or passive dosimeters in order to place them deep inside the CSBF shielding and irradiate them. For this reason, the removable sample holder concrete block provides of 3 slots of 10 cm x 10 cm cross section that are centered along the vertical axis of the block, so that they can be filled with the samples. In figure 2 the recent layout of CSBF is presented when the removable sample holder concrete block is inserted. The position 1 is located at a height of 10.5 cm, measured from the bottom of the removable sample holder concrete block, the position 2 at 85.4 cm height and the position 3 at 160.35 cm height. There is also a possibility of placing samples on the top of the block, mentioned as position 4 at 200 cm height. The block is easily inserted in and extracted from the CSBF shielding in a specifically designed shaft, with dimensions 40 cm x 40 cm x 240 cm. The neutron spectra predicted by the Monte Carlo code FLUKA at the 4 positions of the removable sample holder concrete block are shown in the figure 3.

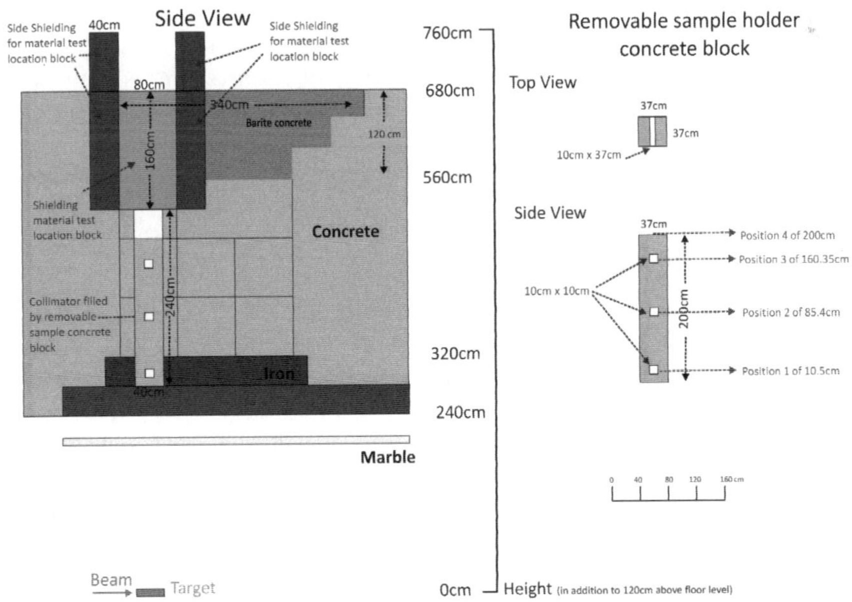

Figure 2. CSBF upgrade layout for measurements with the removable sample holder concrete block inserted in the facility.

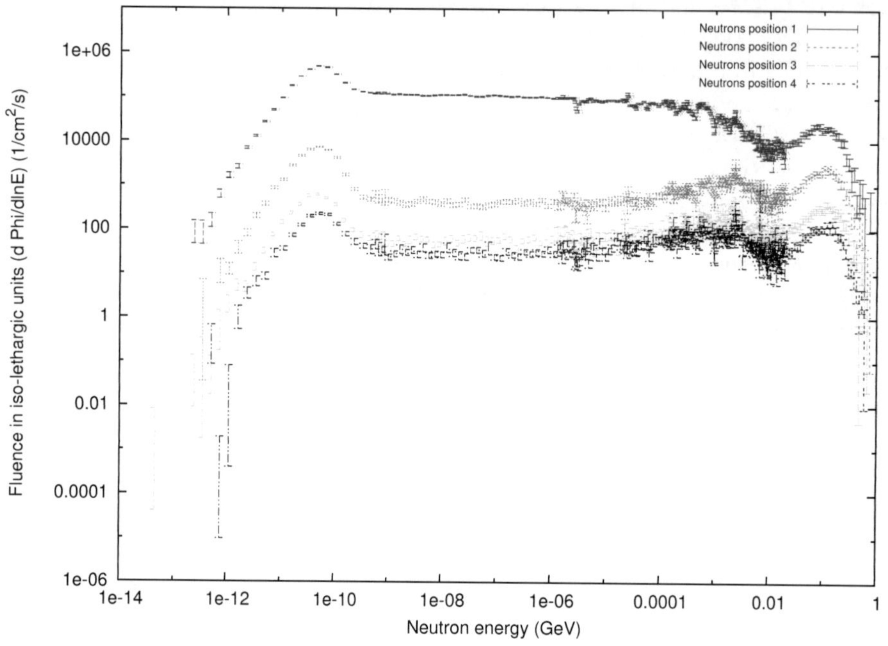

Figure 3. Neutron fluence spectra predicted by FLUKA at the 4 different positions on the removable sample holder concrete block for an average beam intensity of $6.7{\cdot}10^{10}$ protons per second.

2.2. CSBF platform

The CSBF platform was created at 560 cm above beam line level, indicated in figure 2, and allows placing active detectors or dosimeters attached to phantoms on the top of the shielding, for measuring their response to a deep penetration neutron spectrum. In order to use the platform, two barite concrete blocks of 120 cm height have to be removed. The CSBF platform measurements can be performed in parallel to activation measurements at the removable sample holder concrete block activation measurements.

2.3. Shielding material test location

The shielding material test location was designed for measuring the spectrum averaged attenuation length (λ) of various shielding materials (e.g. standard concrete, barite concrete, hematite concrete, colemanite concrete, magnetite concrete and cast iron). The available blocks are of 20, 40 and 80 cm thickness and for each material the ambient dose equivalent rate can be measured up to a shielding thickness corresponding to approximately 4λ or 5λ (depending on the material). The spectrum averaged attenuation length can then be determined by a second ambient dose equivalent rate measurement with an additional shielding layer, ideally with a thickness of the order of 1λ. To perform these measurements the removable sample holder concrete block has to be removed from the shaft so that there is a collimated radiation field heading directly to the shielding material test location. FLUKA simulations have been performed and have shown that the contributions from neutrons scattered on the side walls of the shaft and the surrounding shielding structure are less than 3% of the ambient dose equivalent rate.

3. Beam parameters and configurations

This section presents the beam parameters and the facility configurations that were used during the activation experiments in September 2016. The beam intensity was measured with a Secondary Emission Chamber, whose measurement values are logged in the CERN measurement database. An intensity calibration factor was applied to the counts per pulse to obtain the number of protons per pulse. This calibration factor had been previously obtained with aluminium foil activation measurements using sodium isotopes with a statistical uncertainty of 7% of the γ-spectrometry analysis [14].

A beam size of 1.2 cm x 1.2 cm Full Width at Half Maximum (FWHM) was used for the FLUKA simulations as specified in the layout of the beam line and confirmed by online beam profile measurements [14].

The average beam intensity of CHARM, binned in 5 minutes long intervals, from July 6 to July 12 and from September 16 to September 22, 2016 when the experiments were conducted, is shown in the figure 4 and figure 5 respectively. The first irradiation period corresponds to the irradiation of the aluminum samples, the second to the irradiation of the bismuth samples and the third to the irradiation of the indium samples.

The beam passes through the upstream Proton Irradiation facility (IRRAD) before impacting on the CHARM target. During the period of the experiment, Silicon samples with a total thickness of 0.2 cm were placed into the beam in IRRAD and these samples were also taken properly into account in the FLUKA simulations.

The shielding layout of the CSBF as implemented in the FLUKA simulations is shown in figure 2. The chemical composition of the concrete, the barite concrete and the cast iron implemented in the FLUKA Monte Carlo simulations for the shielding with their respective densities are listed in table 1 and in table 2.

During the activation experiment, the copper target of 8 cm diameter and 50 cm length has been used inside the CHARM facility. Inside the target room, there are four movable shielding walls, each of 20 cm thickness and made out of concrete or iron. They can be placed between the target and the irradiation positions for electronics components inside the CHARM facility in different combinations, so that the irradiation spectra are adjusted to the desired radiation field (energy and intensity) during the tests. The movable shielding walls are also indicated in the figure 1. For this activation experiment, only one configuration of the four movable shielding walls was used during the different irradiation periods, namely all movable shielding walls retracted from the facility. The configuration has been properly taken into account in the FLUKA Monte Carlo simulations.

Table 1. Chemical composition [7] and density of concrete.

Concrete	Density 2.4 g/cm^3		
Element	Weight fraction (%)	Element	Weight fraction (%)
Hydrogen	0.561	Silicon	16.175
Carbon	4.377	Sulfur	0.414
Oxygen	48.204	Potassium	0.833
Sodium	0.446	Calcium	23.929
Magnesium	1.512	Titanium	0.173
Aluminium	2.113	Iron	1.263

Table 2. Chemical composition [15] and density of cast iron [7].

Cast Iron	Density 7.2 g/cm^3
Element	Weight fraction (%)
Iron	92.3
Carbon	3.85
Manganese	0.3
Silicon	3.4
Phosphorus	0.08
Sulfur	0.02
Cobalt	0.05

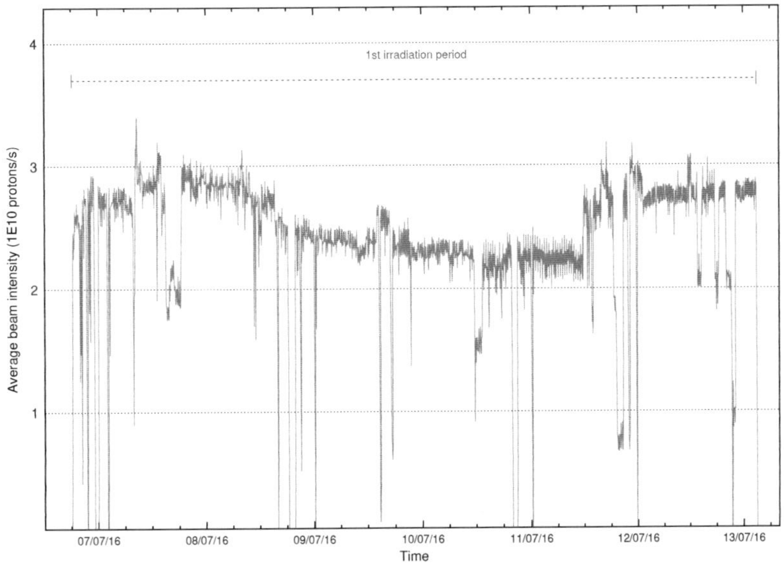

Figure 4. Average beam intensity of the CHARM facility during the activation experiments in July 2016 binned in 5 minutes long intervals.

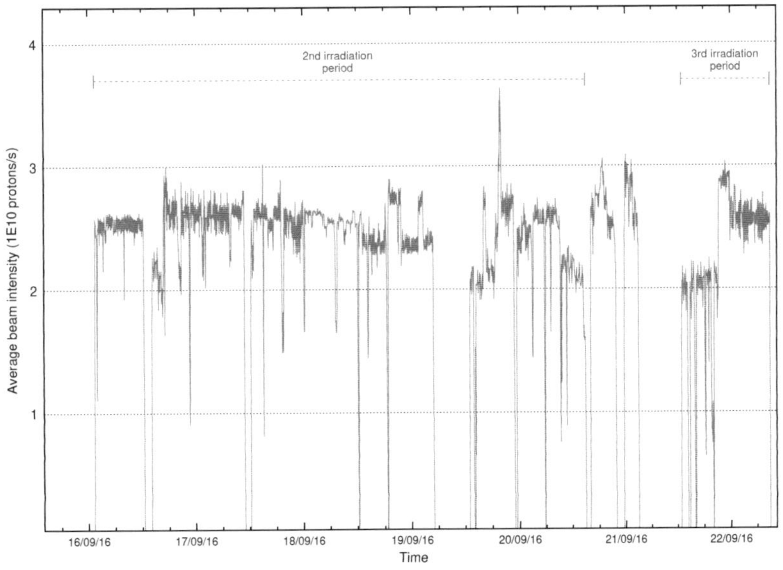

Figure 5. Average beam intensity of the CHARM facility during the activation experiments in September 2016 binned in 5 minutes long intervals.

IOP Conf. Series: Journal of Physics: Conf. Series **1046** (2018) 012004 doi:10.1088/1742-6596/1046/1/012004

4. Activation samples and their irradiation

Fourteen disk samples in total, five bismuth samples, five aluminium samples and four indium samples, have been irradiated. These samples were placed in the removable concrete sample holder block of the CSBF at the 4 positions as indicated in figure 2. All the details of the samples including their location, irradiation time, dimensions, weight, cooling time and duration of γ-spectrometry measurements are presented in table 3.

Table 3. Description of the irradiated samples and details of the γ-spectrometry measurements.

Sample Material	Irradiation Location	Duration of irradiation (h)	Start of irradiation (date and time)	Dimensions (mm)	Weight (g)	Cooling time (h)	γ-spectrometry measurement duration (h)
Al	Position 1	149.3	06/07/2016, 18:12 PM	40(diam.)×3	10.4	7	3
Al	Position 2	149.3	06/07/2016, 18:12 PM	40(diam.)×3	10.4	7	3
Al	Position 3	149.3	06/07/2016, 18:12 PM	40(diam.)×3	10.4	7	3
Al	Position 3	149.3	06/07/2016, 18:12 PM	60(diam.)×15	116.9	7	3
Al	Position 4	149.3	06/07/2016, 18:12 PM	60(diam.)×15	116.9	7	3
Bi	Position 1	109.6	16/09/2016, 01:25 AM	20(diam.)×2	6.23	2	3
Bi	Position 2	109.6	16/09/2016, 01:25 AM	40(diam.)×4	49.5	2 5 26 240	3 12 16 48
Bi	Position 3	109.6	16/09/2016, 01:25 AM	40(diam.)×4	49.52	2 4.5 25.5 187	3 12 16 48
Bi	Position 3	109.6	16/09/2016, 01:25 AM	80(diam.)×10	523.5	2	3
Bi	Position 4	109.6	16/09/2016, 01:25 AM	80(diam.)×10	540.34	2 5 26	3 12 16
In	Position 1	20	21/09/2016, 12:55 PM	20(diam.)×2	4.59	6	8
In	Position 2	20	21/09/2016, 12:55 PM	40(diam.)×4	36.38	6.5	8
In	Position 3	20	21/09/2016, 12:55 PM	80(diam.)×10	375.49	1	3
In	Position 4	20	21/09/2016, 12:55 PM	80(diam.)×10	384.77	1.5	3

IOP Conf. Series: Journal of Physics: Conf. Series **1046** (2018) 012004 doi:10.1088/1742-6596/1046/1/012004

5. Comparison of FLUKA simulation results to measured production yields

The simulation results were obtained by first scoring the neutron fluence spectra with FLUKA. Then, the neutron fluence was folded with cross section data for the bismuth isotopes, ^{24}Na and ^{115m}I [16], shown in figure 6, to obtain the predicted production yields per atom per primary proton on the target.

The activities of the bismuth isotopes,^{24}Na and ^{115m}I were measured for the bismuth, aluminum and indium samples respectively using γ-spectrometry, sometimes even at different cool-down times. In case of multiple samples for the same materials at the same position or multiple γ-spectrometry measurements of the same sample, the activities selected were the ones with the lowest uncertainty of the γ-spectrometry measurements. These activities have been converted to the production yields by taking into account the corresponding irradiation profiles with 5 minutes long binning and the corresponding cool-down times.

The production yields predicted by FLUKA and measured by γ-spectrometry are presented in figure 7 and in table 4. The agreement between FLUKA predictions and γ-spectrometry measurements for the production yields is generally better than a factor of 2.

The contributions that have been taken into account for the uncertainty estimation are shown in table 5. The uncertainty of the beam size has negligible impact on the results as verified by FLUKA simulations. The materials placed in IRRAD during the period of the experiment were taken into account in the simulations. The uncertainty of the production yields coming from the uncertainty of the materials placed in IRRAD is far below 1%. A hypothetical change in the concrete density would provoke a change on the slope of the dependence of the yields on the depth of the shielding plotted in figure 7 and the effect of the change would increase with increased shielding thickness.

The cumulative distribution to the production yields as function of the neutron energy is presented in figure 8 for the sample placed at a concrete shielding thickness of 160.35 cm in the CSBF. From this figure it can be seen that for sodium-24, the neutron energy range contributing to the production yield is quite large whereas for the bismuth isotopes the energy ranges are narrower and located around the respective cross section peaks. The 10%, 25%, 75% and 90% quantiles of the production yield distribution for the various radionuclides, are presented in table 6, quantifying this effect.

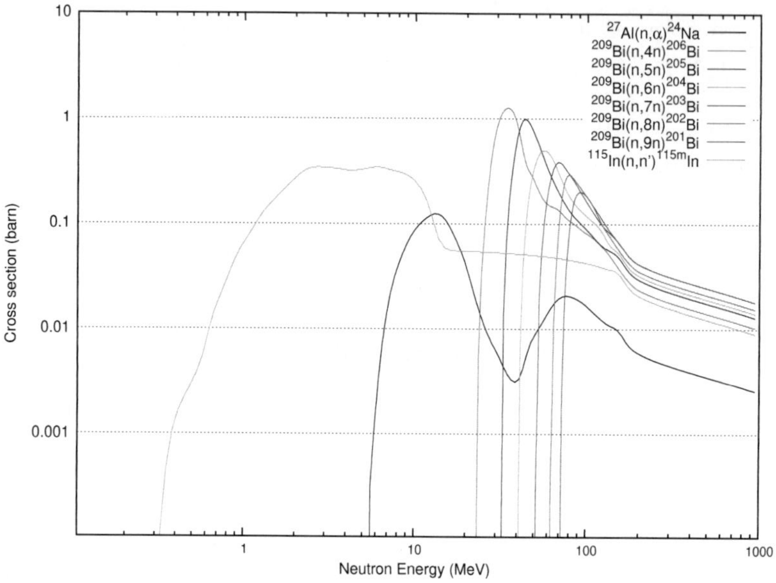

Figure 6. Production cross sections of the bismuth isotopes,^{24}Na and ^{115m}I as a function of the neutron energy [16].

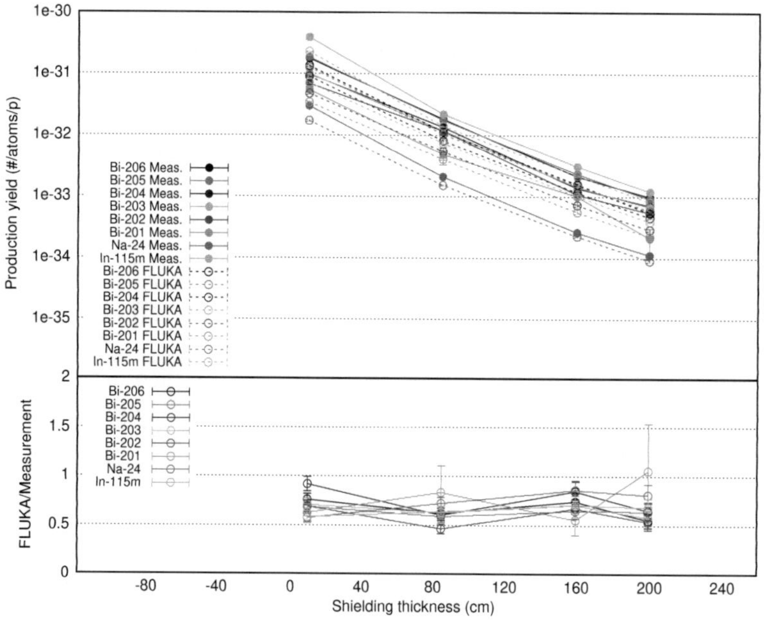

Figure 7. Comparison of the predicted production yields by FLUKA and the measured production yields by γ-spectrometry for bismuth, sodium and indium radionuclides as function of shielding thickness.

Table 4. Predicted production yields by FLUKA and measured production yields by γ-spectrometry.

Radionuclide	Position / Height (cm)	Predicted Production yield by FLUKA (number/atom/p)	Uncertainties FLUKA (%)	Measured Production yield from γ-spect analysis (number/atom/p)	Uncertainties of γ-spect analysis (%)	Ratio Predicted/ Measured	Uncertainties Ratio (%)
Bi-206	1 / 10.5	1.34E-031	±3.48	1.76E-031	±7.14	0.74	±7.94
	2 / 85.4	1.13E-032	±5.56	1.82E-032	±7.25	0.62	±9.14
	3 / 160.35	1.60E-033	±9.50	2.15E-033	±7.50	0.74	±12.11
	4 / 200	5.48E-034	±11.38	9.91E-034	±7.25	0.55	±13.49
Bi-205	1 / 10.5	1.26E-031	±3.47	1.83E-031	±7.58	0.69	±8.33
	2 / 85.4	1.03E-032	±5.28	1.75E-032	±7.92	0.59	±9.52
	3 / 160.35	1.53E-033	±9.19	2.36E-033	±8.72	0.65	±12.67
	4 / 200	5.89E-034	±11.29	9.2E-034	±7.96	0.64	±13.81
Bi-204	1 / 10.5	8.99E-032	±3.40	9.78E-032	±7.25	0.92	±8.01
	2 / 85.4	8.14E-033	±5.16	1.35E-032	±7.20	0.60	±8.86
	3 / 160.35	1.17E-033	±9.09	1.38E-033	±7.50	0.85	±11.79
	4 / 200	4.62E-034	±11.27	7.11E-034	±7.20	0.65	±13.38
Bi-203	1 / 10.5	7.08E-032	±3.59	1.00E-031	±8.78	0.71	±9.48
	2 / 85.4	7.42E-033	±5.21	1.18E-032	±8.54	0.63	±10.01
	3 / 160.35	1.02E-033	±9.15	1.44E-033	±9.76	0.71	±13.38
	4 / 200	3.93E-034	±10.89	6.79E-034	±8.38	0.58	±13.74
Bi-202	1 / 10.5	4.76E-032	±4.00	6.78E-032	±8.01	0.69	±8.96
	2 / 85.4	5.44E-033	±5.35	1.17E-032	±9.09	0.47	±10.55
	3 / 160.35	7.33E-034	±9.47	1.09E-033	±11.32	0.67	±14.76
	4 / 200	2.92E-034	±11.10	5.45E-034	±10.41	0.54	±15.21
Bi-201	1 / 10.5	3.41E-032	±4.19	5.45E-032	±13.72	0.63	±14.35
	2 / 85.4	4.05E-033	±5.42	4.85E-033	±31.88	0.83	±32.34
	3 / 160.35	5.61E-034	±9.72	1.01E-033	±25.77	0.56	±27.54
	4 / 200	2.26E-034	±11.16	2.19E-034	±44.16	1.06	±45.55
Na-24	1 / 10.5	1.71E-032	±2.62	2.98E-032	±7.14	0.57	±7.61
	2 / 85.4	1.52E-033	±5.15	2.10E-033	±7.47	0.72	±9.07
	3 / 160.35	2.28E-034	±9.01	2.65E-034	±7.31	0.86	±11.60
	4 / 200	9.16E-035	±11.01	1.13E-034	±7.96	0.81	±13.59
In-115m	1 / 10.5	2.26E-031	±1.48	3.87E-031	±10.04	0.58	±10.15
	2 / 85.4	1.40E-032	±4.68	2.19E-032	±7.12	0.64	±8.52
	3 / 160.35	2.22E-033	±8.81	3.13E-033	±9.90	0.71	±13.25
	4 / 200	8.26E-034	±10.39	1.21E-033	±8.32	0.68	±13.31

Table 5. Uncertainties taken into account for the uncertainty estimation of the production yields.

	Source of uncertainty	**Uncertainty on production yield**
Simulations	Statistical uncertainty	1.4-5.4 %
	Concrete density*	0.5-10 % **
Measurements	γ-spectrometry	1.3-14 % ***
	sample weights	1 %
	beam intensity	7 %
	beam momentum	< 1 %
	beam position and profile	< 1 %
	target density	< 1 %
	target dimensions	< 1 %

* Uncertainty of the concrete density is 0.05 g/cm^3.
** The concrete density uncertainty leads to an uncertainty of the production yield of 0.5% for 10.5 cm of concrete and of up to 10% for 200 cm of concrete.
*** Except for ^{201}Bi at 85.4 cm, 160.35 cm, 200 cm, see table 4.

IOP Conf. Series: Journal of Physics: Conf. Series **1046** (2018) 012004 doi:10.1088/1742-6596/1046/1/012004

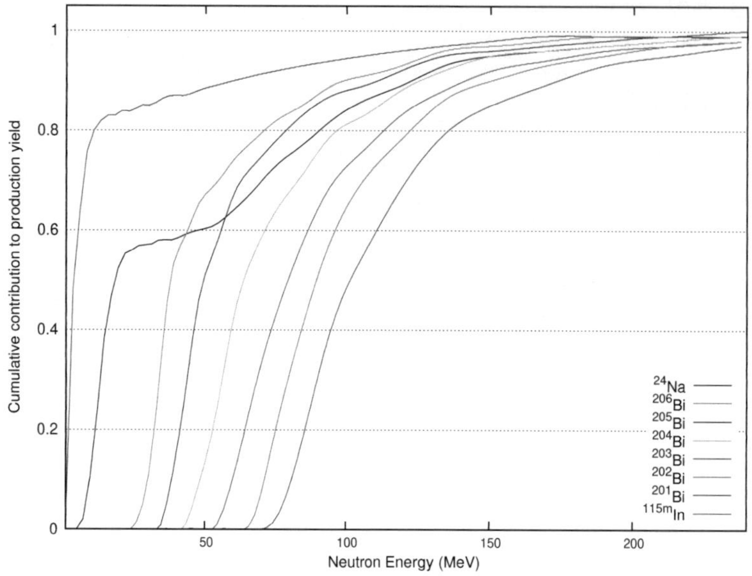

Figure 8. Cumulative contribution to the production yields at a concrete shielding thickness of 160.35 cm as a function of the neutron energy.

Table 6. The 10%, 25%, 75% and 90% quantiles of the production yield distribution for the various radionuclides at a concrete shielding thickness of 160.35 cm.

	Neutron Energy [MeV]			
Radionuclide	$q_{0.1}$	$q_{0.25}$	$q_{0.75}$	$q_{0.9}$
Bi-201	79.9	87.8	128	173
Bi-202	70.6	77.4	114	149
Bi-203	59.7	66.1	104	141
Bi-204	48.3	54.4	89.4	126
Bi-205	38.2	42.5	70.2	110
Bi-206	29.5	32.7	61.5	100
Na-24	9.39	12.2	79.7	122
In-115m	1.63	2.33	7.99	59.7

6. Summary & Conclusions

The CERN High Energy Accelerator Mixed Field facility (CHARM) has been constructed in the CERN PS East Experimental Area. The facility receives a pulsed proton beam from the CERN PS with a beam momentum of 24 GeV/c with $5{\cdot}10^{11}$ protons per pulse with a pulse length of 350 ms. The maximum average beam intensity is $6.7{\cdot}10^{10}$ p/s.

The shielding of the CHARM facility also includes the CERN Shielding Benchmark Facility (CSBF) situated laterally above the target. This facility allows deep-penetration benchmark studies of various shielding materials.

From our experience gained through a previous activation campaign in July 2015, we decided to upgrade the CSBF in 2016 in order to facilitate the procedure of sample placement and to add more functionalities in the facility. Based on the results mentioned above [8, 9], FLUKA was used for the design of the upgrade in CSBF (as a reliable Monte Carlo simulation tool).

An activation foil experiment has been conducted at the upgraded CSBF from September 21 to September 28, 2016. Bismuth, Aluminium and Indium cylindrical samples were placed in the removable concrete sample holder block of the CSBF at different heights. The production rates computed from the activities of the irradiated samples measured by γ-spectrometry have been compared to the estimated production rates from FLUKA Monte Carlo simulations. The agreement is at a level of a factor of 2.

This agreement is good for deep shielding penetration studies and is consistent with previous similar studies at the CERN-EU High Energy Reference Field facility (CERF) [7].

Acknowledgments

We thank our colleagues from the IRRAD and CHARM operation teams for their support and for providing beam time and the CERN Experimental Area group for the integration and construction of the CSBF. We would also like to show our gratitude to the CERN transport group for their assistance. We are also grateful to our colleagues from the CERN γ-spectrometry laboratory for their support.

References

[1] Mekki J, Brugger M, Alia R G, Thornton A, Dos Santos Mota N C and Danzeca S 2016 CHARM: A Mixed Field Facility at CERN for Radiation Tests in Ground, Atmospheric, Space and Accelerator Representative Environments, *IEEE TRANSACTIONS ON NUCLEAR SCIENCE*, **VOL. 63, NO. 4** p 2106-2114

[2] Froeschl R 2014 Radiation Protection Assessment of the Proton Irradiation facility and the CHARM facility in the East Area, Tech. Rep. CERN-RP-2014-008-REPORTS-TN, EDMS 1355933

[3] Froeschl R, Brugger M, Roesler S 2014 The CERN High Energy Accelerator Mixed Field (CHARM) facility in the CERN PS East Experimental Area *Proceedings of SATIF12, NEA/NSC/R(2015)3, Batavia, Illinois, United States* p 14-25

[4] Adorisio C and Roesler S 2011 Attenuation of high-energy neutrons in different shielding materials. A FLUKA benchmark for an intercomparison study, Tech. Rep. CERN-DGS-2011-035-RP-TN, EDMS 1139144

[5] Agosteo S, Birattari C, Para A F et al. 2001 Neutron measurements around a beam dump bombarded by high energy protons and lead ions *Nuclear Instruments and Methods in Physics Research A* **459** p 58-65

[6] Agosteo S, Pozzi F, Silari M et al. 2013 Attenuation in iron of neutrons produced by 120 GeV/c positive hadrons on a thick copper target *Nuclear Instruments and Methods in Physics Research B* **312** p 36-41

[7] Nakao N, Taniguchi S, Roesler S et al. 2008 Measurement and calculation of high-energy neutron spectra behind shielding at the CERF 120 GeV/c hadron beam facility *Nuclear Instruments and Methods in Physics Research B* **266** p 93-106

[8] Iliopoulou E et al. 2015 Measurements and FLUKA Simulations of Bismuth and Aluminium Activation at the CSBF, Tech. Rep. CERN-RP-2015-118-REPORTS-TN, EDMS 1566978

[9] Iliopoulou E et al. 2016 Measurements and FLUKA Simulations of Bismuth and Aluminum Activation at the CERN Shielding Benchmark Facility (CSBF) *Proceedings of SATIF 13, Dresden, Germany*

[10] Iliopoulou E et al. 2017 Measurements and FLUKA Simulations of Bismuth and Aluminum Activation at the CERN Shielding Benchmark Facility (CSBF) *submitted to Nuclear Instruments and Methods in Physics Research A*

[11] Böhlen T T et al. 2014 The FLUKA Code: Developments and Challenges for High Energy and Medical Applications, *Nuclear Data Sheets* **120** p 211-214

[12] Fassò A, Ferrari A, Ranft J and Sala P R 2005 FLUKA: a multi-particle transport code, Tech. Rep. CERN-2005-10 (2005), INFN/TC-05/11,SLAC-R-773

[13] Froeschl R et al. 2016 Radiation Protection Aspects of the Commissioning and Operation of the CHARM facility *Proceedings of SATIF 13, Dresden, Germany*

[14] Curioni A, Froeschl R, Glaser M, Iliopoulou E, La Torre F P, Pozzi F, Ravotti F and Silari M 2017 Single- and multi-foils ^{27}Al (p,3pn) ^{24}Na activation technique for monitoring the intensity of high-energy beams *Nuclear Instruments and Methods in Physics Research A* **858** p 101-105

[15] Lazzaroni M 2015 Personal Communication

[16] Maekawa F et al, 2001 Production of a Dosimetry Cross Section Set Up to 50 MeV *Proc. 10th International Symposium on Reactor Dosimetry, Sep. 12-17, 1999, Osaka, Japan, American Society for Testing and Materials* p 417

IOP Conf. Series: Journal of Physics: Conf. Series **1046** (2018) 012005 doi:10.1088/1742-6596/1046/1/012005

Sensitivity of ambient dose equivalent to the concentration of cobalt impurity present in stainless steel

N. Shetty, V. Olsovcova and R. Versaci
ELI Beamlines, Za Radnici 835, Dolni Brezany, Czech Republic
E-mail: nikhil.shetty@eli-beams.eu

Abstract. Stainless steels contain nickel in large amounts (about 8 %) to improve its corrosion and heat resistance. Traces of cobalt are present in nickel, which are hard to separate because of its chemical similarity. Therefore, cobalt content in steel is restricted to a maximum of 2 parts per mille for applications in nuclear industry, as natural cobalt (composed of 100% Co-59) transmutes into highly radioactive Co-60 by absorbing a thermal neutron. Co-60 has a rather long half-life of 5.3 years decaying to stable Ni-60 by emitting 2 gammas of 1.17 MeV and 1.33 MeV during the process. These hard gammas will be mostly responsible for the dose rates seen in the next few tens of years. Therefore, it is important to consider the activation of cobalt in steel and estimate the dose contributed by it. Monte Carlo simulations are performed where stainless steel samples with different cobalt concentrations are irradiated with thermal and epithermal neutrons. The ambient dose equivalent, H*(10), from irradiated samples is found to be linearly proportional to the concentration of cobalt. This paper explains the motivation, the procedure, and the detailed results of the simulations.

1. Introduction

ELI Beamlines is a high-power laser facility under development, where laser-plasma interactions are expected to produce secondary sources of radiation. The radiation protection group at ELI Beamlines assesses prompt radiation and activation of materials. In order to plan radioactive material handling, activation studies are routinely performed using Fluka [1, 2] simulations for the most commonly used materials. The material composition is obtained from the manufacturers, however it does not account for the contamination. Contamination of stainless steel with cobalt is a major issue in radiation protection, and this paper presents the studies done on the sensitivity of radiation dose to the concentration of cobalt contamination.

Stainless steel is used ubiquitously in particle accelerators for beam pipes, interaction chambers, vacuum chambers, structural materials, nuts, bolts, screws, etc. As an example, figure 1 shows the E2 experimental room at ELI Beamlines. The room hosts the beam transport and interaction chambers that are made of stainless steel.

Stainless steels contain nickel in large amounts (about 8 %) to improve its corrosion and heat resistance. Traces of cobalt are present in nickel, which are hard to separate because of their chemical similarity. Therefore, cobalt content in steel is restricted to a maximum of 2 parts per mille for applications in nuclear industry as natural cobalt (composed of 100 % cobalt-59) transmutes into highly radioactive cobalt-60 by absorbing a thermal neutron. Cobalt-60 has a

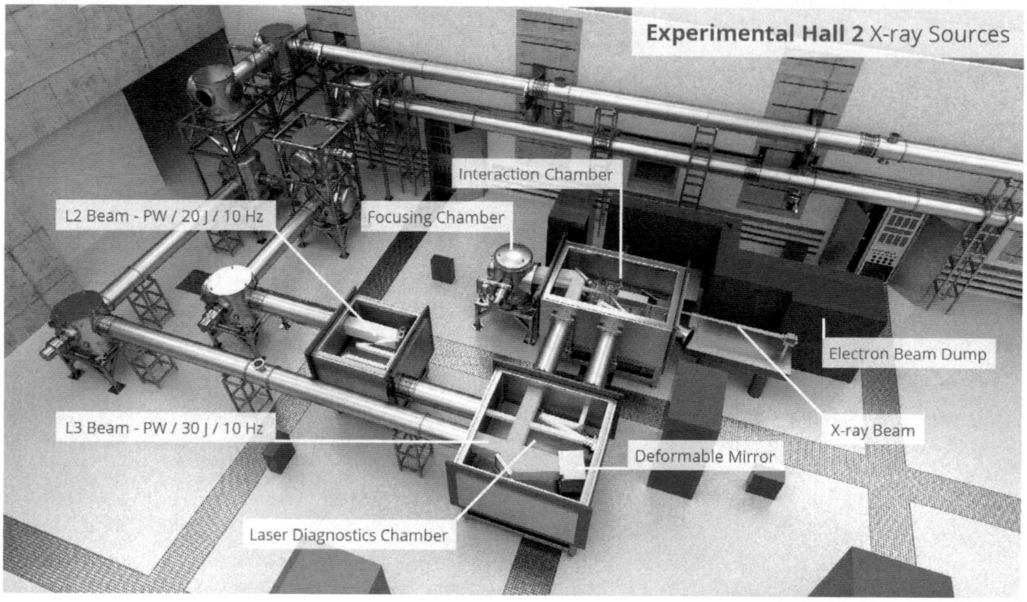

Figure 1. Engineering rendering of E2 experimental room at ELI Beamlines that is hosting laser beam transport and interaction structures made from steel.

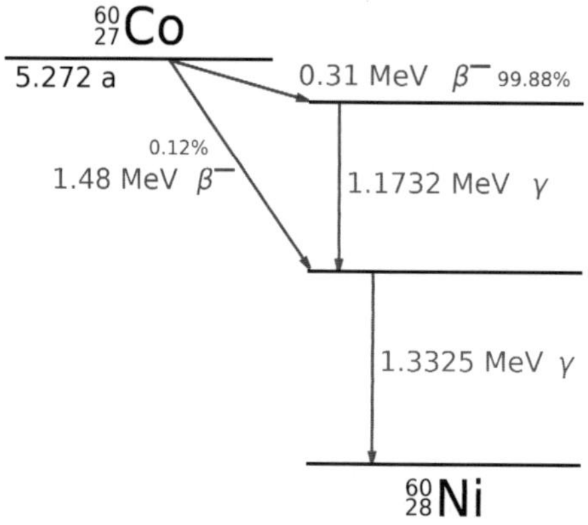

Figure 2. Decay scheme of cobalt-60.

rather long half-life of 5.3 years decaying to stable nickel-60 by emitting 2 gammas of 1.17 MeV and 1.33 MeV during the process (see figure 2). These hard gammas will be mostly responsible for the dose rates seen in the next few tens of years. Therefore, it is important to consider the activation of cobalt in stainless steel, and estimate the dose contributed by it.

Co-59 has a significant thermal neutron capture cross section. It also has a large resonance for neutron capture around 100 eV as shown in figure 3.

2. Source term and simulation of induced activation

Laser-plasma interactions at ELI Beamlines are expected to produce protons up to 250 MeV and electrons up to 10 GeV. There are other sources of radiation such as X-rays and ions, however these are not important as the former for activation. The intensities of protons and electrons

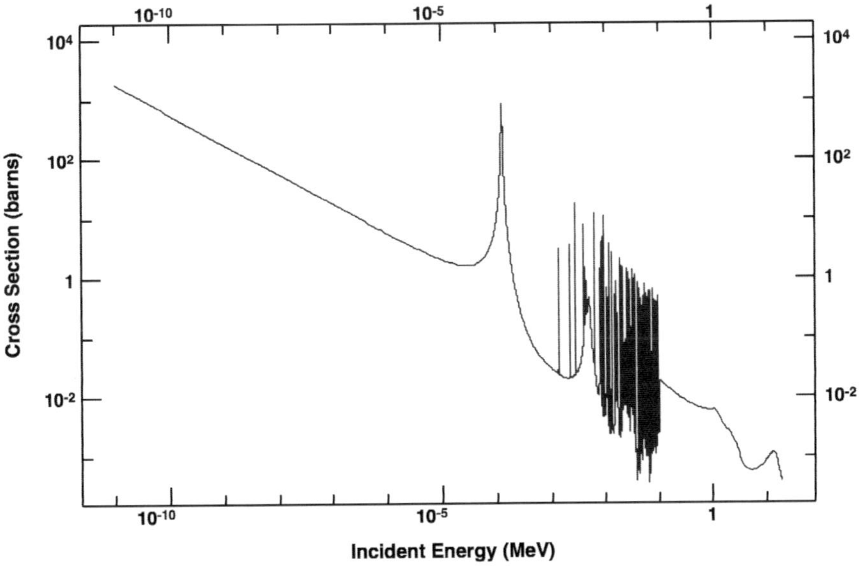

Figure 3. Co-59(n,gamma)Co-60 reaction cross section as in ENDF [3].

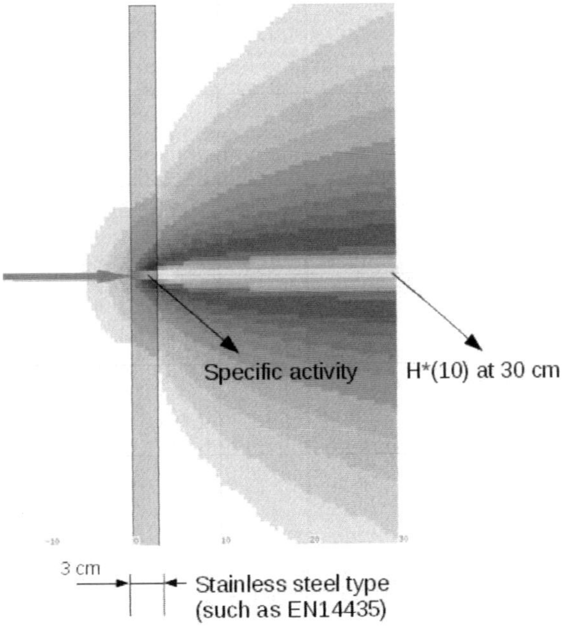

Figure 4. Simulation scheme for activation studies.

can reach up to 10^{10} particles per shot, with a repetition rate up to about 10 Hz.

For structural material activation studies, a 3 cm thick, very large plane of the material is irradiated with a pencil beam of particles. The ambient dose equivalent, H*(10) at 30 cm from the material is scored, along with the nuclide inventory (see figure 4 obtained via Flair [4]).

Figure 5 shows the neutron spectra generated by proton and electron projectiles, respectively

IOP Conf. Series: Journal of Physics: Conf. Series **1046** (2018) 012005 doi:10.1088/1742-6596/1046/1/012005

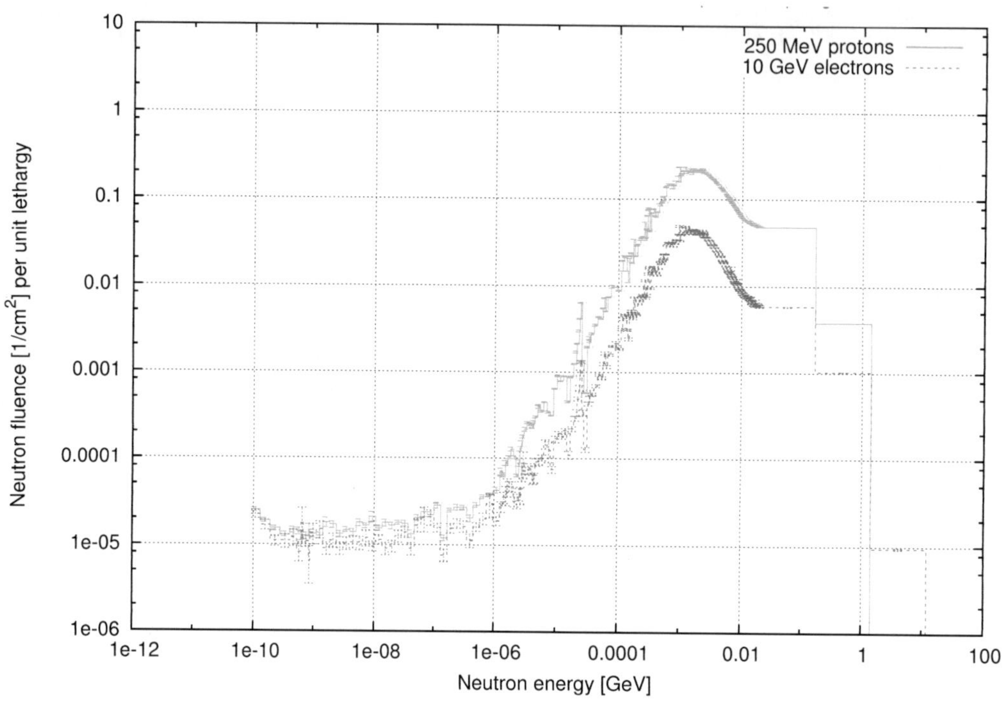

Figure 5. Neutron spectra from a 3 cm thick EN14435 type stainless steel target due to 250 MeV protons and 10 GeV electrons.

of 250 MeV and 10 GeV. The upper limit of neutron energy is up to about a few tens of MeV, and the lower limit depends on the thermalization. Thermalization is offered by the thick concrete walls surrounding the neutron source, and the amount of activation of cobalt is directly proportional to the extent of thermalization, as the neutron absorption probability is higher in the thermal region. Therefore the simulations are always run with the concrete walls (cage) present.

The yearly irradiation profile has been assumed as follows. For the first 20 days of the month, there are 4 hours of irradiation each day. The beam is off for the rest of the month. This pattern is repeated for 12 months. The dose and nuclide inventory are scored at the following cooling times: 0 second, 1 day, 1 week, 4 weeks and 26 weeks, after the end of the irradiation.

3. Results

Figure 6 shows the dose rate decay curve for 250 MeV proton irradiation on stainless steel with an intensity of 10^{10} particles per second with and without contamination. Immediately after the end of irradiation, the dose rate rises by up to about 20% for steel contaminated with 10 parts per mille of cobalt when compared to dose rate in uncontaminated steel. For other cooling times, the increase in dose rate is up to about 10%, as can be seen in figure 7. The difference in dose rate increase between short and long cooling times can be attributed to the presence of short-lived isotopes such as meta-stable isomers of cobalt-60.

Table 1 shows the concentration of cobalt contamination and the specific activity of cobalt-60. There exits a linear relationship between the two. Hence, and unsurprisingly, in the long-term, the dose is linearly proportional to the concentration of cobalt. Note that the statistical error of the calculated activity is negligible.

IOP Conf. Series: Journal of Physics: Conf. Series **1046** (2018) 012005 doi:10.1088/1742-6596/1046/1/012005

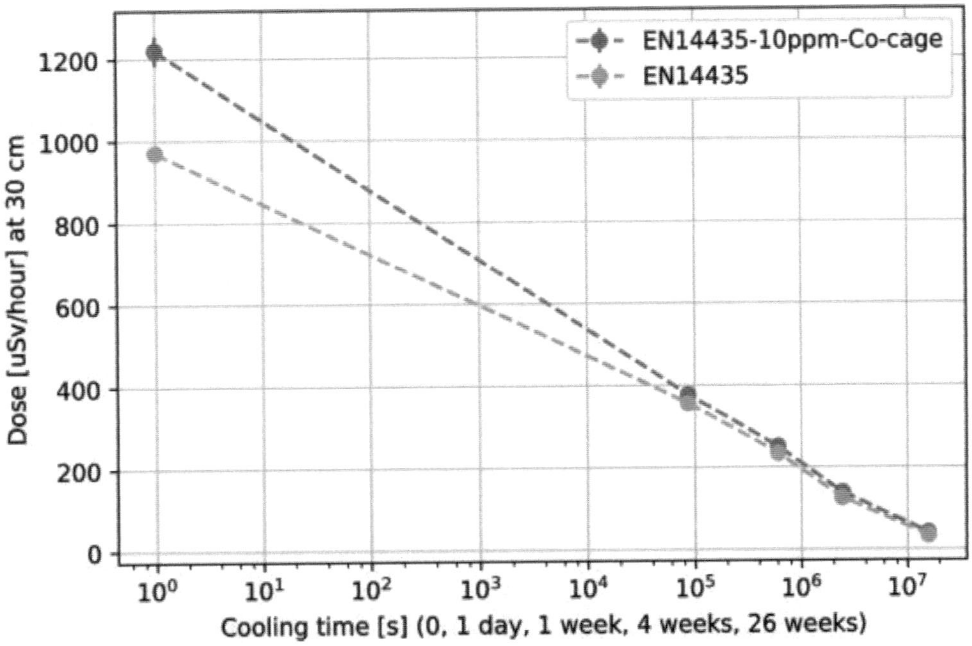

Figure 6. Comparison of dose rate decay curves for uncontaminated steel and steel contaminated with 10 parts per mille of cobalt. Error bars are smaller than the markers

Table 1. Linear relationship between cobalt-59 contamination and specific activity of cobalt-60.

Co-59 ppm	Co-60 Bq/cm^3 (end of irradiation)
0	3.975
2	8.924
10	26.03
20	44.34
40	74.94

4. Conclusions

Natural cobalt contamination in stainless steel is a real problem faced by every particle accelerator laboratories producing neutrons. The paper presented the sensitivity studies performed for activation of cobalt contamination in stainless steel. Even small amounts (2 parts per mille) of cobalt can lead to a systematic increase in the observed dose by up to about 20%, immediately after shutdown, and about 10% thereafter. Hence uncertainty or lack of knowledge of cobalt contamination in steel can lead to a non-negligible systematic error in the observed dose.

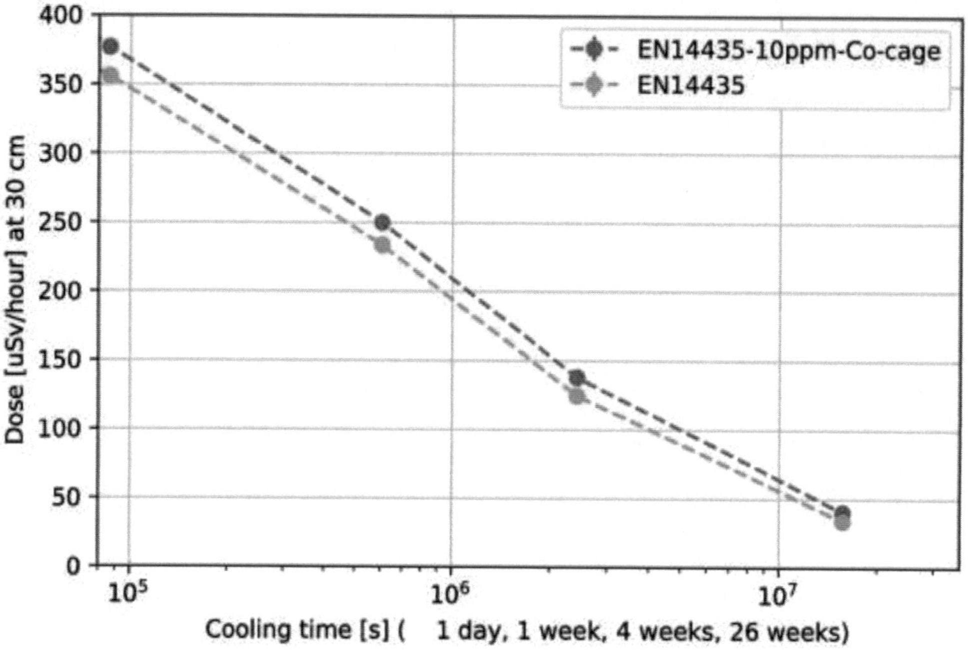

Figure 7. Comparison of dose rate decay curves for longer cooling times. Error bars are smaller than the markers.

Acknowledgments

Acknowledgments Supported by the project ELI - Extreme Light Infrastructure - phase 2 (CZ.02.1.01/0.0/0.0/15_008/0000162) from European Regional Development Fund.

References

[1] Bohlen T T, Cerutti F, Chin M P W, Fasso' A, Ferrari A, Ortega P G, Mairani A, Sala P R, Smirnov G and Vlachoudis V 2014 *Nuclear Data Sheets* B **120** 211-214
[2] Ferrari A, Sala P R, Fasso' A and Ranft J 2005 *CERN-2005-10 (2005)*
[3] Chadwick M B et al 2011 *Nucl. Data Sheets* **112** 2887-2996
[4] Vlachoudis V 2009 *Proc. Int. Conf. on Mathematics, Computational Methods and Reactor Physics (MC 2009)*

IOP Conf. Series: Journal of Physics: Conf. Series **1046** (2018) 012006 doi:10.1088/1742-6596/1046/1/012006

A method for radiological characterization based on fluence conversion coefficients

Robert Froeschl

CERN, 1211 Geneva 23, Switzerland

E-mail: robert.froeschl@cern.ch

Abstract. Radiological characterization of components in accelerator environments is often required to ensure adequate radiation protection during maintenance, transport and handling as well as for the selection of the proper disposal pathway. The relevant quantities are typical the weighted sums of specific activities with radionuclide-specific weighting coefficients. Traditional methods based on Monte Carlo simulations are radionuclide creation-event based or the particle fluences in the regions of interest are scored and then off-line weighted with radionuclide production cross sections. The presented method bases the radiological characterization on a set of fluence conversion coefficients. For a given irradiation profile and cool-down time, radionuclide production cross-sections, material composition and radionuclide-specific weighting coefficients, a set of particle type and energy dependent fluence conversion coefficients is computed. These fluence conversion coefficients can then be used in a Monte Carlo transport code to perform on-line weighting to directly obtain the desired radiological characterization, either by using built-in multiplier features such as in the PHITS code or by writing a dedicated user routine such as for the FLUKA code. The presented method has been validated against the standard event-based methods directly available in Monte Carlo transport codes.

1. Introduction

The radiological characterization of components after their use in an accelerator facility is an important task for operational radiation protection, due to the need to define adequate radiation protection measures during maintenance, transport or handling of a component or due to the need to select a proper disposal pathway.

The relevant radiological hazard factor for radiological characterization can often be written as a weighted sum, with radionuclide specific weighting coefficients, of specific activities.

The commonly used methods for obtaining the required specific activities are briefly presented and analyzed. This analysis highlights the need of an alternative method providing fast convergence, automatic normalization and good visualization capabilities.

A new method based on fluence conversion coefficients is developed that satisfies these requirements. The formulation of the method and its implementation is presented and its capabilities are demonstrated on selected examples.

Finally, promising future applications of this method are discussed.

2. Radiological characterization

The radiological hazard factor, denoted H, relevant for radiological characterization can often be written as a weighted sum, with radionuclide specific weighting coefficients w_b, of specific activities A_b:

$$H = \sum_b \frac{1}{w_b} A_b = \sum_b \frac{1}{w_b} \sum_r \sum_e T_{br} \, P_{re} \, m_e \tag{1}$$

The mass specific activity A_b of radionuclide b can often be written in terms of the production rate P_{re} of radionuclide r from the chemical element e

$$P_{re} = \frac{N_A}{M_e} \sum_{i=\mathrm{p,n},\pi^+,\pi^-} \int \phi_i(E) \, \sigma_{i,e,r}(E) \, dE, \tag{2}$$

where N_A is Avogadro's constant and M_e is the atomic weight of the chemical element e. The sum is typically extended over protons (p), neutrons (n) and charged pions (π^+, π^-) for high energy hadron accelerators. In this formulation, the natural isotope abundances for each element e are taken into account, i.e. the cross section $\sigma_{i,e,r}(E)$ is an abundance weighted average of the cross sections of each isotope of element e. Furthermore, $\Phi_i(E)$ denotes the radiation fluence for the various secondary particles ($i = \mathrm{p}, \mathrm{n}, \pi^+, \pi^-$). The expression $\sum_e P_{re} \, m_e$ corresponds to the production rate of isotope r in the whole component.

The time evolution of the specific activity of radionuclide b, i.e. the build–up of all radionuclides r and the full decay chain leading to isotope b, can be described by the matrix T_{br}, involving the time-dependent beam intensity profile and typically expressed using the *bateman coefficient* [1, 2].

2.1. Standard methods for radiological characterization

Monte Carlo radiation transport simulation codes can be used to compute the radiological hazard factor H needed for radiological characterization. There are two standard methods for these calculations that are often employed when using Monte Carlo simulation codes. These two methods are presented in the next paragraphs and their specific use in the FLUKA [3, 4] and PHITS [5, 6, 7] Monte Carlo radiation transport simulation codes are discussed.

Event based: The first standard method to obtain the required specific activities is based on the events of creation of radionuclides during interactions in the radiation transport.
In FLUKA, the RESNUCLE card provides this capability, either scoring production yields or directly activities for a given irradiation profile. This scoring is based on regions defined in the geometry of the simulation.
In PHITS, the T-Yield tally can score production yields, also based on regions defined in the geometry of the simulation. The time evolution to obtain specific activities has to be performed offline.

Fluence spectra based: The second standard method to obtain the required specific activities is based on the calculation of fluence spectra in the Monte Carlo simulation and then weighting these spectra with radionuclide production cross-sections in a post-processing step. The time evolution is also accounted for in the post-processing step.
In FLUKA, the USRTRACK card provides this capability whereas in PHITS the T-Track tally can be used for this purpose.
The post-processing can be done with various codes, e.g. JEREMY [8, 9] or ActiWiz [10, 11].

A detailed comparison of the these two types of standard methods is presented in table 1. This comparison clearly demonstrates the need for an alternative method with **fast convergence**, **automatic normalization** and **good visualization capabilities** for a large number of cells.

Table 1. Comparison of the two standard methods for radiological characterization, described in section 2.1 and the fluence conversion coefficients method, described in section 3.

Method	Pros	Cons
Event based	Built-in capability with reduced human error probability Very good for small number of regions close to main interaction zone Uses cross-sections from MC code Material compositions from MC input	Convergence can be slow for regions away from main interaction zone Region-based: Mass normalization has to be done *manually* Difficult visualization Cumbersome for large number of regions
Fluence spectra based	Better convergence properties Possibility to evaluate alternative material compositions	Cross-sections decoupled from MC code Region-based: Volume normalization has to be done *manually* Difficult visualization Cumbersome for large number of regions Higher post-processing needs
Fluence conv. coeff. based	Fully integrated into existing tool-chain Very good visualization of larger number of regions Automatic mass normalization Better convergence properties	Cross-sections decoupled from MC code Higher pre-processing needs

3. Fluence conversion coefficients method

The radiological hazard factor H in equation 1 can be expressed in terms of fluence conversion coefficients, denoted FCCs. This can be achieved by inserting equation 2 into equation 1, approximating the integrals over the particle energies by discrete sums and then re-ordering the summation in a way to move the particle-type summation and the energy summation as well as the particle fluences to the front.

$$\begin{aligned} H &= \sum_b \frac{1}{w_b} \sum_{r,e} T_{br}\, P_{re}\, m_e \\ &\approx \sum_b \frac{1}{w_b} \sum_{r,e} T_{br} \left(\frac{N_A}{M_e} \sum_{i=\mathrm{p,n},\pi^+,\pi^-} \sum_E \Phi_{i,E}\, \sigma_{i,e,r}(E) \right) m_e \\ &= \sum_{i=\mathrm{p,n},\pi^+,\pi^-} \sum_E \Phi_{i,E} \underbrace{\left(\sum_{b,r,e} \frac{1}{w_b} T_{br} \frac{N_A}{M_e} \sigma_{i,e,r}(E)\, m_e \right)}_{K_{i,E}} \end{aligned} \tag{3}$$

Equation 3 now expresses the radiological hazard factor H in terms of fluence conversion coefficients $K_{i,E}$, yielding also the definition for the $K_{i,E}$ coefficients. The basic idea of the presented method is to precompute the fluence conversion coefficients $K_{i,E}$ for a given radiological characterization task and then to apply them on-line during the Monte Carlo radiation transport simulation.

IOP Conf. Series: Journal of Physics: Conf. Series **1046** (2018) 012006 doi:10.1088/1742-6596/1046/1/012006

3.1. Implementation
A dedicate code for the pre-computation of the desired fluence conversion coefficients has been developed. This code takes the material composition, the irradiation and cool-down profile as well as the radionuclide specific weight coefficients for the hazard factor as input from the user. Based on these data and the cross-section data set of the JEREMY code [8], it computes the fluence conversion coefficients as defined in equation 3 and stores them either as F77 source code snippet for a dedicated FLUKA user routine or as PHITS multipliers in the PHITS input file format.

The code is written in the python programming language [12] and therefore automatically exposes a scripting interface. It has a very compact code base, consisting of 300 lines of code plus 1000 lines of code for the time evolution that is based on the JEREMY code, with decay data from JEFF 3.1.1 [13]. However, it is designed so that it can be easily interfaced with other cross-section data sets or time evolution engines.

The application of the fluence conversion coefficients during the Monte Carlo radiation transport simulation is achieved in FLUKA by a dedicated `fluscw` user routine that is then applied to a USRBIN scoring. In PHITS, the multipliers can be applied to a T-Yield tally. In both codes, the scoring of the radiological hazard factor is performed on a mesh geometry and is therefore not directly coupled to the region definitions of the simulation geometry.

4. Validation
A simulation of a proton beam hitting a copper target has been performed to verify the correctness of the fluence conversion coefficients method with respect to the FLUKA code. The copper target is a cylinder with a radius of 5 cm and a length of 80 cm. The target has been segmented in the simulation geometry into 4 parts of 20 cm length for scoring purposes. The proton beam with a momentum of 1 GeV/c is impinging onto the target at the center of the circular front face. The target is surrounded by a cylindrical-symmetric structure of air, concrete, cast iron and concrete. The layout of the simulation geometry is shown in figure 1.

Several radiological quantities have been scored with FLUKA as well as with the fluence conversion coefficients method in the copper target and the shielding structure. These quantities comprise saturation activities of selected radionuclides, their activities after certain irradiation and cool-down periods and multiples of the Swiss Exemption Limit, that is the sum of the specific activities weighted by radionuclide-specific exemption limits defined in the Swiss Radiation Protection Ordinance [14]. A comparison for these quantities is presented in table 2. The agreement is very good and differences larger than the statistical uncertainties can arise from the difference in the underlying radionuclide cross-section set in the used version of FLUKA (FLUKA2011 2c.5) and the ones used to compute the fluence conversion coefficients.

5. Examples and future applications
The capabilities of the fluence conversion coefficients method are demonstrated with several examples. This is followed by the discussion of promising future applications of the fluence conversion coefficients method.

5.1. Examples
The first example is the calculation of the multiples of the Swiss Exemption Limit for concrete after 10 years of irradiation followed by 2 years of cool-down for the validation geometry (see section 4) with the PHITS code. The results are shown in figure 2 for a beam intensity of 1 proton per second on the target.

The following examples show multiples of the Swiss Exemption Limit after 10 years of irradiation followed by 2 years for different materials for the CHARM facility [15] that is located in the CERN PS East Experimental Area. It receives a primary proton beam from the CERN

IOP Conf. Series: Journal of Physics: Conf. Series **1046** (2018) 012006 doi:10.1088/1742-6596/1046/1/012006

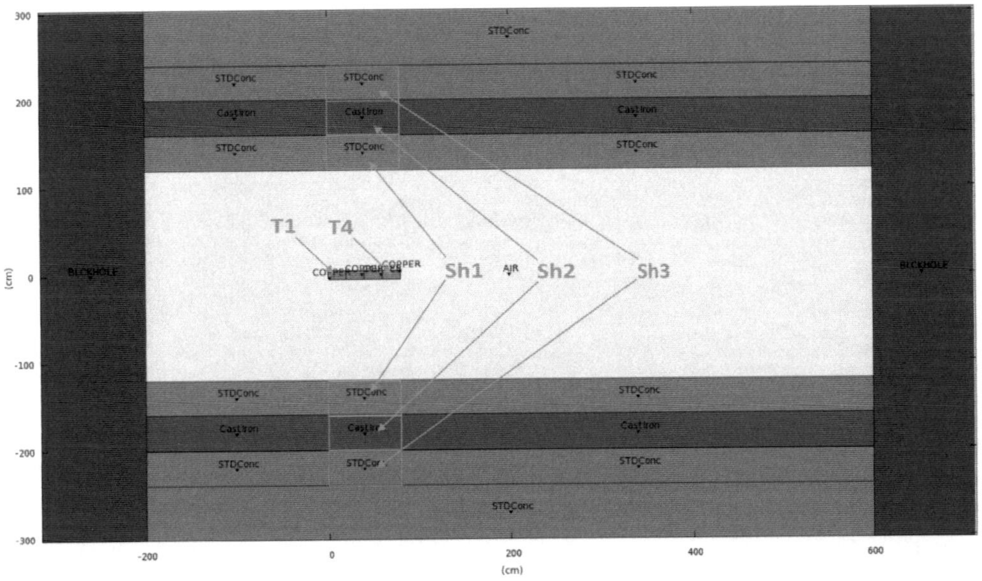

Figure 1. Cut through the cylindrical-symmetric simulation geometry used for the validation of the fluence conversion coefficients method with respect to the FLUKA code. The cylindrical-symmetric scoring volumes are indicated by the orange arrows.

Table 2. Comparison of several radiological quantities scored with FLUKA as well as with the fluence conversion coefficients method for validation purposes.

Radiological Quantity		Material	Region	FLUKA*	FFCs*	FLUKA/FCCs
Co-60	Production Yield	Copper	T1	1.98e-06±0.8%	1.536e-06±0.1%	1.29 ± 0.81%
			T4	5.82e-09±14%	4.897e-09±2.1%	1.19 ± 14%
Na-24	Production Yield	Concrete	Sh1	2.91e-09±3.3%	2.668e-09±0.4%	1.09 ± 3.3%
			Sh3	8.02e-11±16%	8.680e-11±2%	0.924 ± 16%
	Specific Activity (10y, 2d)	Concrete	Sh1	3.15E-10±3.3%	2.885e-10±0.4%	1.09 ± 3.3%
			Sh3	8.67E-12±16%	9.385e-12±2%	0.924 ± 16%
Na-22	Specific Activity (10y, 2y)	Concrete	Sh1	1.20E-10±2.8%	1.443e-10±0.3%	0.832 ± 2.8%
			Sh3	2.25E-12±16%	3.251e-12±1.8%	0.692 ± 16%
Mn-54	Specific Activity (10y, 2y)	Cast Iron	Sh2	1.78e-10±1.5%	1.641e-10±1.7%	1.08 ± 2.3%
Multiple of Swiss Exemption Limit (10y, 2y)		Concrete	Sh1	1.48e-10±3.1%	1.539e-10±0.2%	0.962 ± 3.1%
			Sh3	4.16e-12±5.7%	4.575e-12±0.9%	0.909 ± 5.8%
		Cast Iron	Sh2	2.37e-10±2.9%	2.339e-10±0.5%	1.01 ± 2.9%

* Production Yields in [nucl/primary]
Specific Activities in [(Bq/g)/(primary/s)]
Multiple of Swiss Exemption Limit in [1/(primary/s)]

PS with a beam momentum of 24 GeV/c that impinges on a copper or aluminum target. The beam is slowly extracted with a maximum pulse length of 350 ms and a maximum average design beam intensity of $6.7 \cdot 10^{10}$ protons per second. The layout of the CHARM facility is shown in figure 3 and a discussion of the radiation protection aspects of the CHARM facility can be found in [16].

The multiples of the Swiss Exemption Limit after 10 years of irradiation with $6.7 \cdot 10^{10}$ protons per second followed by 2 years of cool-down is shown for concrete in figure 4 and for cast iron in figure 5. These results can be used in the future during the decommissioning phase to establish a zoning that identifies shielding structures that are expect to be activated below the Swiss Exemption Limits and that are therefore candidates for clearance from regulatory control.

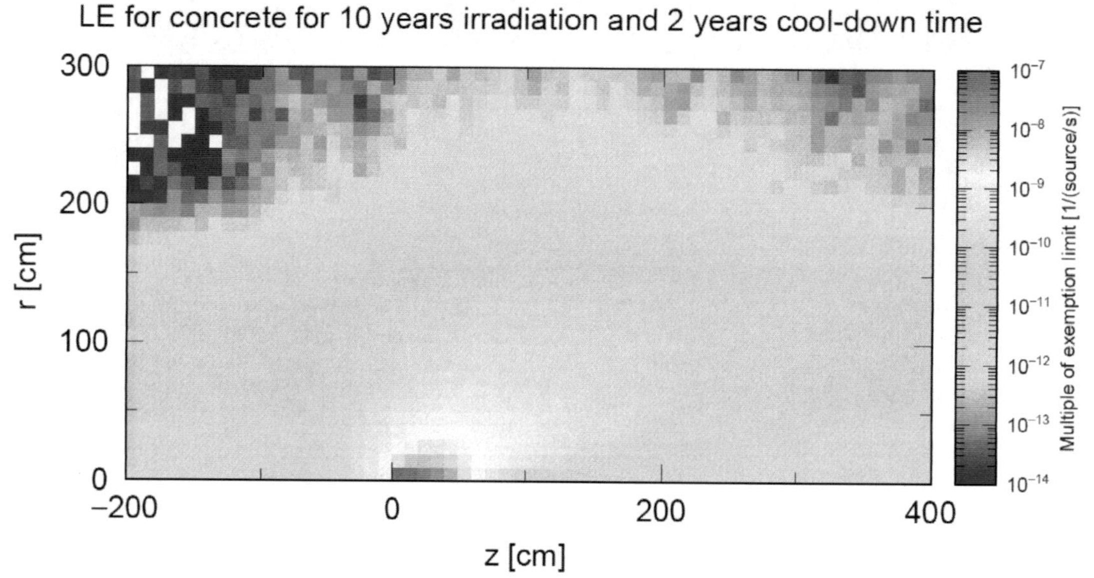

Figure 2. Multiples of the Swiss Exemption Limit computed with the PHITS code for concrete after 10 years of irradiation followed by 2 years of cool-down for the validation setup described in section 4.

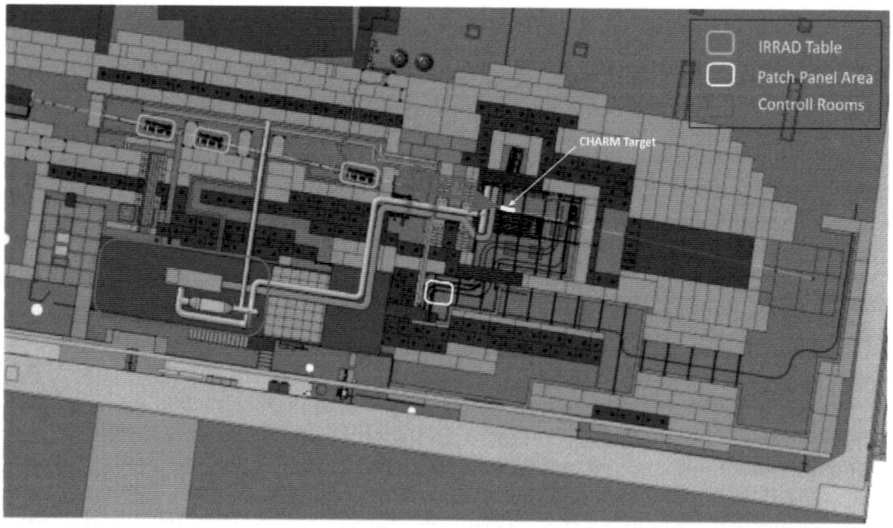

Figure 3. Integration drawing of the CHARM facility in the CERN PS East Experimental Area

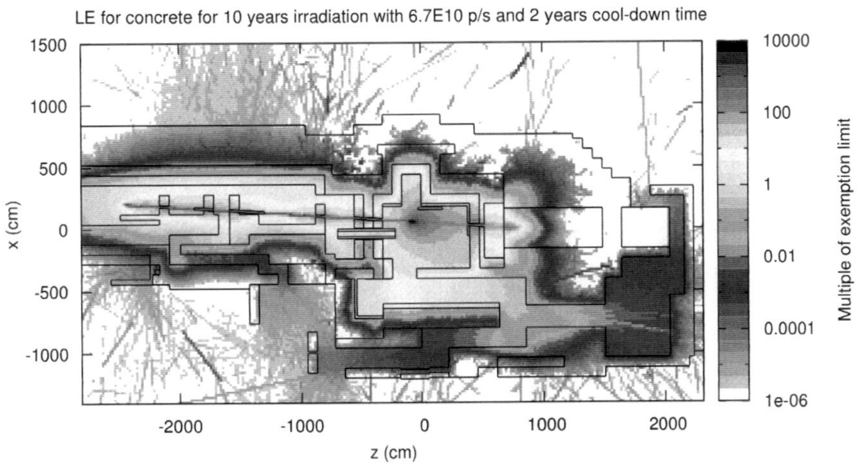

Figure 4. Multiples of the Swiss Exemption Limit at beam line height computed with the FLUKA code for concrete after 10 years of irradiation followed by 2 years of cool-down for the CHARM facility described in section 5.1.

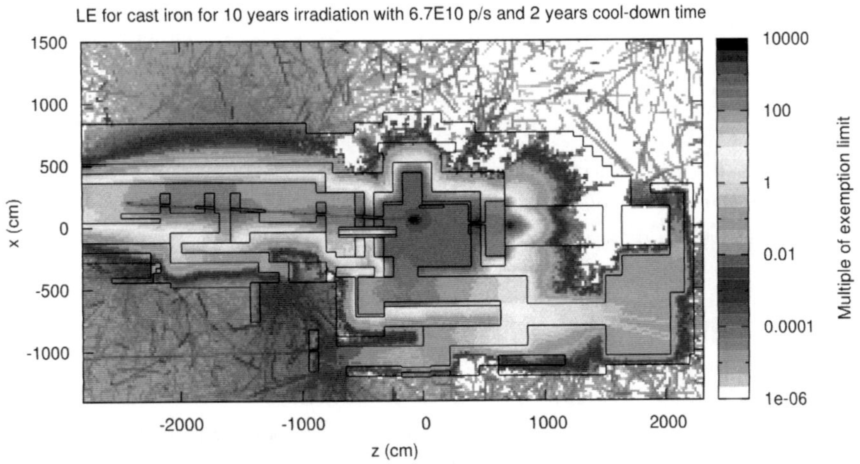

Figure 5. Multiples of the Swiss Exemption Limit at beam line height computed with the FLUKA code for cast iron after 10 years of irradiation followed by 2 years of cool-down for the CHARM facility described in section 5.1.

5.2. Future applications

The fluence conversion coefficients method is very general in its nature. These are some of the promising future fields of application of this method:

Radioactive waste pre-characterization: The shielding structures for new accelerator facilities are often designed using Monte Carlo radiation transport simulations. The radiological characterization of these structure, based on the fluence conversion coefficients method, can be already performed under conservative assumptions for the exploitation parameters. Estimates for the radioactive waste quantities as well as costs for its disposal can then be computed based on the obtained radiological characterization results.

Clearance zoning: It is often practical to establish a zoning for maintenance or decommissioning works to delimit the areas where waste might be cleared from regulatory control from areas where the waste will have to be treated as radioactive waste. This type of zoning might be established using the fluence conversion coefficients method for a quite extensive family of waste components.

Earth activation: The prediction of the activation of earth around accelerator tunnels might be required for radiation protection aspects of civil engineering works close to the tunnels or to prove the compliance with activation limits derived from water regulation. The fluence conversion coefficients method might be employed in both assessments.

Planning of activation experiments: One recurring task for the planning of activation experiments is the optimal selection of the masses for activation detectors for different materials, reaction channels and detector placement locations.
A map of the spatial distribution of the saturation activities for a given material and reaction channel might be used to derive the optimal activation detector mass at various locations. This has to be done by taking also the irradiation and cool-down profile as well as the response of the gamma-spectrometry setup used for the activity measurements of the activation detectors after irradiation into account.
An example of a saturation activity distribution map is shown for the production of Na-24 in aluminum in the CHARM facility (described in section 5.1) in figure 6.

6. Summary

The fluence conversion coefficients method for radiological characterization has been presented based on the analysis of the capabilities of the standard methods for radiological characterization.

It is based on the reformulation of the often required weighted sum, with radionuclide specific weighting coefficients, of specific activities, also denoted as radiological hazard factor, as summation of the particle type dependent and energy dependent fluences with a set of fluence conversion coefficients. These fluence conversion coefficients can be precomputed for a given radiological characterization task and then applied on-line during the Monte Carlo radiation transport simulation. This method has been developed for both the FLUKA and PHITS Monte Carlo radiation transport codes.

The fluence conversion coefficients method is complementary to the standard methods for radiological characterization. Its main advantages are fast convergence, automatic mass normalization and good visualization capabilities for a large number of cells. The scoring of the radiological hazard factor can be performed on a mesh geometry and is therefore not directly coupled to the region definitions of the simulation geometry.

The correctness of the fluence conversion coefficients method and its implementation has been verified with respect to the FLUKA code. The capabilities of the fluence conversion coefficients method have been demonstrated by the example and promising future applications have been pointed out.

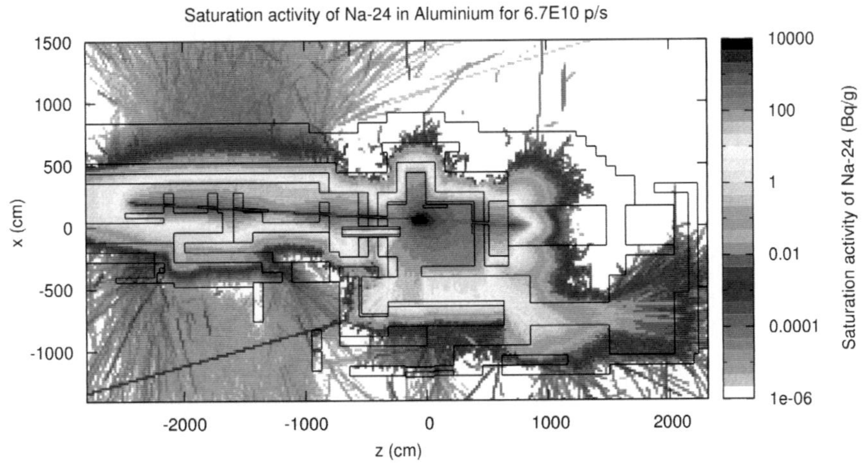

Figure 6. Saturation activity distribution of the production of Na-24 in aluminum at beam line height in the CHARM facility described in section 5.1.

The fluence conversion coefficients method provides very flexible, powerful and extensible tools for radiological characterization at accelerator facilities using Monte Carlo radiation transport simulations.

Acknowledgments

I would like to thank Elpida Iliopoulou for her comments on the manuscript.

References

[1] Bateman H 1910 *Proc. Cambridge Philos. Soc.* **15** 423–427
[2] Shultis J K and Faw R E 2002 *Fundamentals of nuclear science and engineering* (Abingdon: Dekker)
[3] Böhlen TT *et al* 2014 *Nuclear Data Sheets* **120** 211–214
[4] Fassò A, Ferrari A, Ranft J and Sala P R 2005 *FLUKA: a multi-particle transport code* Tech. Rep. CERN-2005-10, INFN/TC-05/11, SLAC-R-773
[5] Sato T, Niita K, Matsuda N, Hashimoto S, Iwamoto Y, Noda S, Ogawa T, Iwase H, Nakashima H, Fukahori T, Okumura K, Kai T, Chiba S, Furuta T and Sihver L 2013 *Particle and Heavy Ion Transport Code System PHITS, Version 2.52* j. Nucl. Sci. Technol. 50:9, 913-923
[6] Niita K *et al. PHITS: Particle and Heavy Ion Transport code System* JAEA-Data/Code 2010-022 (2010)
[7] Cugnon J, Boudard A, Leray S and Mancusi D 2011 *J. Korean Phys. Soc.* **59** 955
[8] Froeschl R, Magistris M and Pereira F 2011 JEREMY, a code for the radiological characterization of accelerator components including detailed uncertainty estimation. presented at the 2nd workshop on Accelerator Radiation Induced Activation(ARIA), Jerusalem, Israel
[9] Froeschl R, Magistris M, Pereira F and Theis C 2014 Computation of radioactivity in particle accelerators and propagation of uncertainties with the JEREMY code, CERN Technical Note CERN-RP-2014-065-REPORTS-TN
[10] Theis C and Vincke H 2016 ActiWiz 3 an overview of the concepts, architecture & new features, CERN Technical Note CERN-RP-2016-117-REPORTS-TN
[11] Theis C and Vincke H 2017 ActiWiz 3 - Getting ready for an activ(ated) future, Presentation given at

the 13th international meeting on nuclear applications of accelerators, AccApp17, 31.7.2017 - 4.08.2017, Quebec, Canada

[12] Rossum G V 2003 *The Python Language Reference Manual* (Network Theory Ltd.) URL `http://docs.python.org`

[13] Santamarina A *et al.* 2009 *The JEFF-3.1.1 Nuclear Data Library* Tech. Rep. JEFF Report 22 Nuclear Energy Agency Paris

[14] Swiss Federal Council 2014 *Swiss Radiological Protection Ordinance (RPO)*, 22 June 1994, Status as of 1 January 2014, Reference 841.501

[15] Mekki J, Brugger M, Alia R G, Thornton A, Mota N C D S and Danzeca S 2016 *IEEE Transactions on Nuclear Science* **63** 2106–2114

[16] Froeschl R, Brugger M and Roesler S 2014 *Proceedings of SATIF12, NEA/NSC/R(2015)3, Batavia,Illinois, United States* 14–25

ActiWiz 3 – an overview of the latest developments and their application

H Vincke[1] and C Theis[1]

[1] CERN, Geneva 23, CH-1211 Geneva, Switzerland

christian.theis@cern.ch

Abstract. In 2011 the ActiWiz code was developed at CERN in order to optimize the choice of materials for accelerator equipment from a radiological point of view. Since then the code has been extended to allow for calculating complete nuclide inventories and provide evaluations with respect to radiotoxicity, inhalation doses, etc. Until now the software included only pre-defined radiation environments for CERN's high-energy proton accelerators which were based on FLUKA Monte Carlo calculations. Eventually the decision was taken to invest into a major revamping of the code. Starting with version 3 the software is not limited anymore to pre-defined radiation fields but within a few seconds it can also treat arbitrary environments of which fluence spectra are available. This has become possible due to the use of ~100 CPU years' worth of FLUKA Monte Carlo simulations as well as the JEFF cross-section library for neutrons < 20 MeV. Eventually the latest code version allowed for the efficient inclusion of 42 additional radiation environments of the LHC experiments as well as considerably more flexibility in view of characterizing also waste from CERN's Large Electron Positron collider (LEP). New fully integrated analysis functionalities like automatic evaluation of difficult-to-measure nuclides, rapid assessment of the temporal evolution of quantities like radiotoxicity or dose-rates, etc. make the software a powerful tool for characterization complementary to general purpose MC codes like FLUKA. In this paper an overview of the capabilities will be given using recent examples from the domain of waste characterization as well as operational radiation protection.

1. Introduction

The ActiWiz code [1] was originally developed to optimize the selection of materials used at high-energy proton particle accelerators, with respect to their radiological hazard. In the second version, features were added which allowed for calculating and analyzing nuclide inventories either for irradiation scenarios included in ActiWiz or based on nuclide production terms originating from external sources [2].

All scenarios which were included in ActiWiz version 2.x (see Ref. [1] for details) are based on a large amount of FLUKA [3][4]simulations which were used to characterize the radiation fields and calculate the associated nuclide production terms. Eventually it was decided to undertake the effort to develop a completely new generation of ActiWiz, which would use arbitrary particle fluence spectra as input and subsequently could calculate the nuclide production terms by itself without the need for further Monte Carlo calculations. In a first step this allowed for efficient extension of the included radiation fields to the 4 major LHC experiments and more flexibility with respect to including future installations.

The calculations are carried out in a deterministic table-driven approach using specific pre-calculated nuclear libraries. In order to analytically calculate the production rate of a radionuclide, the energy as well as particle dependent production cross-section has to be known. Convolution of this function with particle fluence data allows for predicting the number of certain radionuclides produced in a specific radiation field. The disadvantage of this approach is that exhaustive knowledge of all possible isotope production channels is required in order to obtain a complete nuclide inventory. While this is feasible to some extent for some light target elements it becomes impractical and even impossible for compound materials. This problem is generally overcome by Monte Carlo codes using parametrized isotope production models included in their generalized intra-nuclear cascade descriptions. The obvious advantage is that the particle fluence data are calculated in parallel to the isotope production. However, depending on the problem set calculation times might be very long and impractical if radionuclide production in gaseous media or very thin material layers is to be studied. Especially in the domain of radiological characterization these aspects become relevant in order to verify if attention to coatings has to be paid from the radiotoxic point of view.

These are the cases where ActiWiz 3 can be used to complement Monte Carlo studies. In contrast to standard cross-sections it uses tables of nuclide production data which depend on particle type as well as energy. For protons, charged pions, photons as well as neutrons above 20 MeV these tables have been produced by spending ~100 CPU years' worth of FLUKA calculations bombarding thin targets of 85 different chemical elements. The data utilized for neutrons < 20 MeV are derived entirely from the evaluated JEFF cross-section library (version 3.1.1) [5] and include in addition energy dependent production data of isomeric states. The details of the preparation of these data libraries, containing ~17 million records, are beyond the scope of this paper and can be found in detail in Ref [6]. In contrast to ActiWiz 2 the new code now covers the complete periodic system of stable elements as well as a few radioisotopes as possible constituents of compound materials.

This hybrid approach of using parametrized Monte Carlo based data as well as evaluated libraries resolves the shortcoming of having to provide explicit cross-section data as it implicitly includes all reaction channels predicted by a FLUKA simulation + the evaluated data from JEFF. Another advantage is that assessments with high statistical significance can be obtained in very short time, allowing for varying a multitude of parameters like material compositions with different trace elements or irradiation and cool-down scenarios. ActiWiz 3 is based on parallelized algorithms and allows for calculating the complete nuclide production data for 85 target elements within 2 seconds on a standard desktop computer (i7-2600 quad-core at 3.4 GHz).

However, it should be clearly stated that information on the particle fluence is the basic building block which is most often determined via Monte Carlo simulations. Therefore, it should be stressed that the code is not intending to replace Monte Carlo calculations but rather complements them specifically in view of radiological characterization. For this purpose a large number of out-of-the-box analysis functionalities and reports have been implemented, some of which will be discussed in the following sections.

2. Exemplary applications

2.1. Radiological characterization

While nuclide production data are the basis for any RP relevant assessment, eventually it is the nuclide inventory after a specific irradiation and cooling pattern that is of primary interest. For this purpose ActiWiz contains a built-up and decay engine operating at 512 bits floating point accuracy to avoid numerical pitfalls that can arise during the solution of the Bateman equations [2].

In general such a nuclide inventory needs to be put into relation with limits (e.g. clearance limits) or conversion functions (e.g. dose equivalent, inhalation dose, etc.) in order to become meaningful. For this purpose ActiWiz 3 already comes with a large set of such limits/conversion functions like Swiss, Austrian, Euratom, IAEA, US or Japanese clearance limits, and inhalation or ingestion dose conversion functions based on ICRP 72. This allows for easy assessment of for example the radiotoxicity of a material by calculating the sum of the activity with respect to the respective limit over all radionuclides. By default ActiWiz delivers reports relating nuclide activities in relation to any of the include limits/conversion functions. In addition to calculating hazard related quantities like the integral radiotoxicity, also the respective top contributing isotopes are determined.

2.1.1. Radiological relevance of thin coating layers

A typical case where the application of deterministic codes is advantageous over stochastic methods is the assessment of the radiotoxicity of thin layers. In the super conducting cavities of the Large Electron Positron collider (LEP SC cavities) the copper cavities contained a ~2 micro-meter thick Niobium layer. In order to remove such material from regulatory control one needs to assess the radiotoxic relevance of the Niobium layer which, due to its small thickness, becomes very challenging with Monte Carlo methods. This served as a practical example to demonstrate ActiWiz' capabilities for characterization. For this purpose a rough FLUKA model of a cavity assembly was built as depicted in Figure 1.

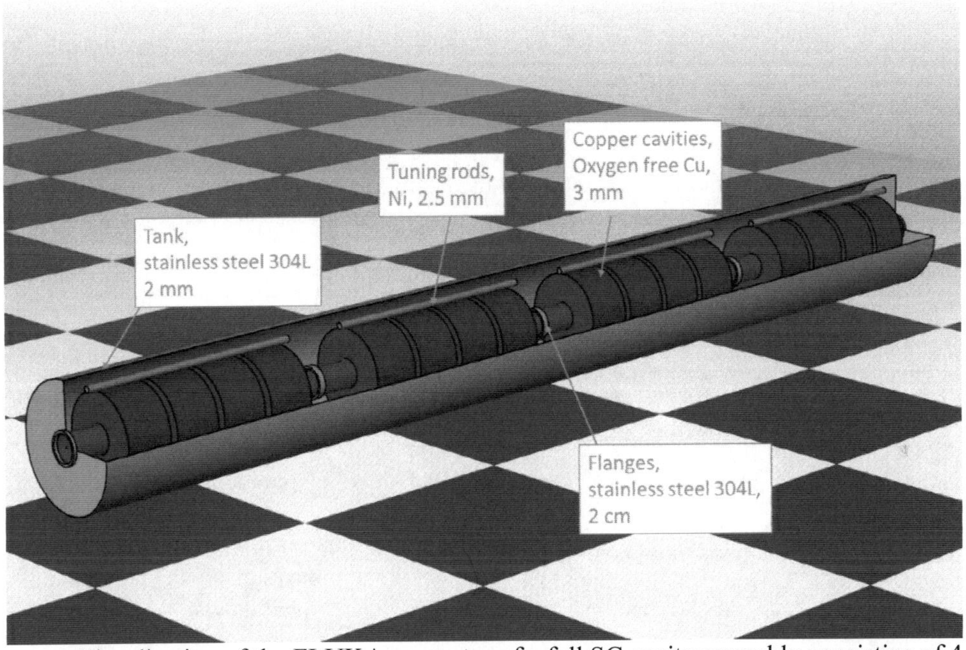

Figure 1 Visualization of the FLUKA geometry of a full SC cavity assembly consisting of 4 cavities with 4 cells each.

It should be noted that the geometry did not include the Niobium coating but only the copper cavities. Using FLUKA the impact of 105 GeV/c electrons at the entrance tube of an assembly was simulated and the ensuing particle spectra in the Copper tube at the beam impact as well as further down-stream have been recorded (see Figure 2 and Figure 3).

IOP Conf. Series: Journal of Physics: Conf. Series **1046** (2018) 012007 doi:10.1088/1742-6596/1046/1/012007

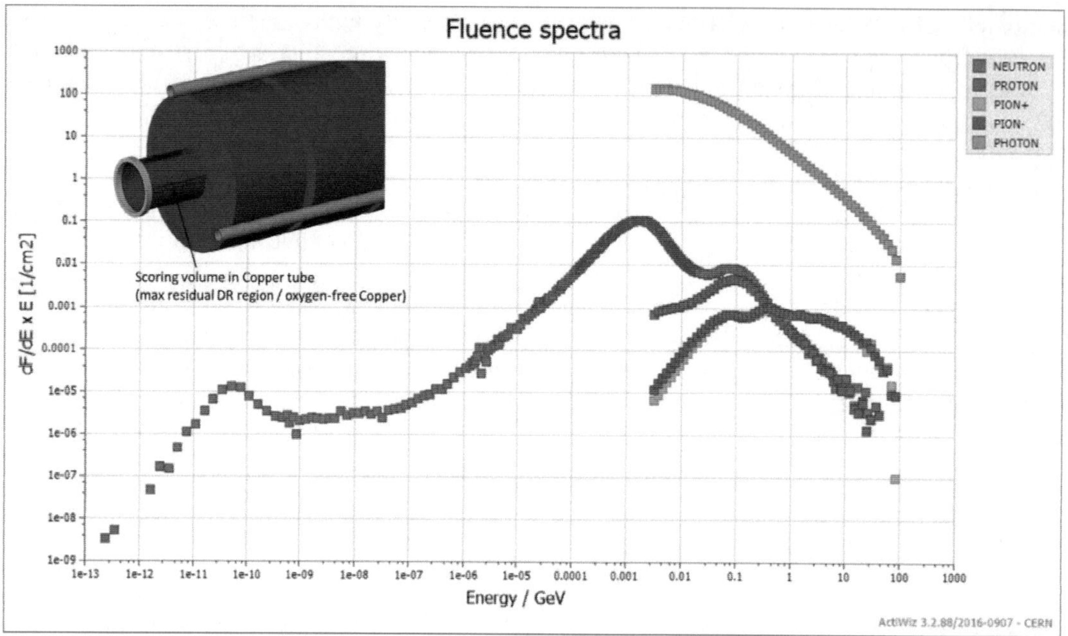

Figure 2 Particle fluence spectra in the Copper tube at the beam impact point. Error bars are generally smaller than 1%.

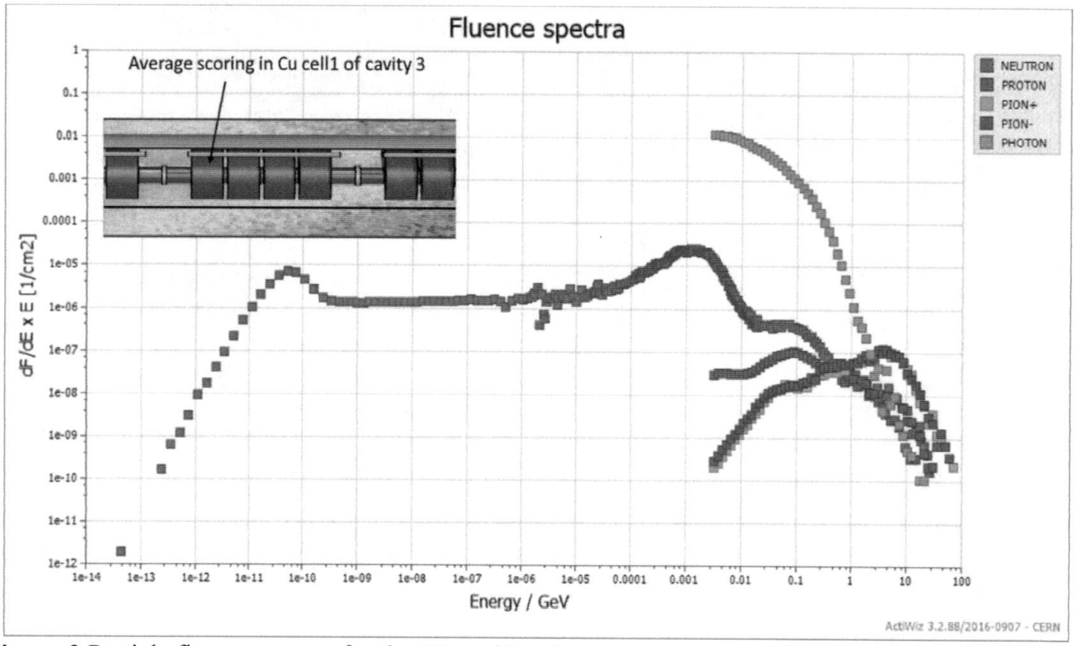

Figure 3 Particle fluence spectra for the "Cu-cell" volume downstream of the beam impact point. Error bars are generally smaller than 1%.

Using the fluence spectra shown above the nuclide inventory for Copper OFE as well as Niobium have been calculated assuming 5 years of irradiation and 15 years of cool-down. Table 1 shows the obtained sum of all nuclide activities over their respective clearance values using different sets of limits.

Table 1 Ratio of the radiotoxicity of Copper vs. Niobium for the different positions as well as different clearance limits.

	Copper vs. Niobium		
	CERN design clearance limit	LE – Swiss clearance limit	LL – future Swiss clearance limit (draft)
Beam impact : Copper vs Niobium	14	7	13
Downstream: Copper vs Niobium	3	3	3

It can be seen that the radiotoxicity of the Copper always outweighs the one of Niobium by at least a factor of 3 up to a factor of 14. Therefore, one can deduce that the Niobium layer can be neglected in any further assessment with respect to the clearance of the Copper, notwithstanding that the total mass of Niobium is negligible in comparison to the base material Copper.

2.1.2. Criticality of chemical constituents

In addition to an isotope based assessment ActiWiz also analysis the contribution to the respective hazard (clearance threshold, dose rate, etc.) by the different chemical constituents of a chemical compound. This is possible as the code provides information on the proportion of how much each chemical element contributes to the production of a specific isotope. Future versions of the code will also provide details differentiating between direct production as well as production via decay chains.

One example where this analysis functionality becomes useful is the characterization of stainless steel. Activation levels of this material after cool-down periods of several months are mainly driven by the amount of Cobalt impurities. However, this information is often unknown, which might pose a considerable problem for any further radiological characterization.

For the purpose of illustration we return to the example of the LEP SC cavities studying the nuclide inventory of stainless steel 304L (assuming 0.1 weight % of Co impurities) at the entrance flange of a cavity assembly as well as the outer cavity tank (see Figure 1). In this context the point of interest was not the radiotoxicity but rather the chemical element mainly responsible for the production of Co-60.

Table 2 Analysis of the dominant radionuclides found with the used stainless steel composition at different locations ("Flange" and "Tank"). In addition, the respective chemical elements are stated which give rise to the production of these radioisotopes [7].

Flange	**Rel. contribution to the radiotoxicity based on CERN's design limits**	**Source**
Co-60	45%	COBALT: 3.96%, IRON: 0.01%, NICKEL: 96.03%
Fe-55	28%	COBALT: 0.01%, IRON: 98.09%, MANGANESE: 0.01%, NICKEL: 1.89%
Tank	**Rel. contribution to the radiotoxicity based on CERN's design limits**	**Source**
Co-60	91%	COBALT: 94.55%, NICKEL: 5.45%
Fe-55	9%	IRON: 99.63%, NICKEL: 0.37%

As can be seen from Table 2 in the vicinity of the beam impact point ("Flange") or along the beam impact trajectory the actual Cobalt content of stainless steel plays a negligible role as the production of

Co-60 originates from Nickel. In contrast to Cobalt, Nickel has a rather well defined content (11 – 15%) as it is an intentional additive in stainless steel.

2.1.3. Criticality of difficult-to-measure (DTM) nuclides
The radiotoxicity of a specific material is defined as the total sum of the activity of each radioisotope with respect to its clearance limit. However, not all radioisotopes can be measured easily, either because they are pure alpha or beta emitters (e.g. Po-210) or have gamma lines which have a very low intensity (e.g. Fe-55). In order to evaluate the fraction of difficult-to-measure (DTM) isotopes contributing to a specific hazard quantity, like radiotoxicity or inhalation dose etc., ActiWiz automatically provides a detailed reporting function. For this purpose the nuclide inventory is processed and all isotopes are listed which are either pure alpha/beta emitters or have gamma lines only which are below 20 keV and/or if the integral intensity of the complete gamma spectrum is below 10^{-4}. A typical example of such a report is given in **Error! Reference source not found.**.

Table 3 Exemplary alpha/beta analysis (excerpt only!) of a nuclide inventory of Indium, irradiated close to the beam-pipe in the Large Hadron Collider for 180 days and 1 hour of cooling.

Difficult to measure nuclides (DTM) Selected hazard: Radiotoxicty based on CERN's design limits				
Id	rel. contribution to hazard [%]	$t_{1/2}$ [s]	Classification	Decay info
H-3	3.19000E-05	3.89E+08	pure alpha/beta	100% β^-
Be-10	1.21000E-10	5.05E+13	pure alpha/beta	100% β^-
C-14	1.34000E-07	1.80E+11	pure alpha/beta	100% β^-
Si-31	1.15000E-06	9432	integral gamma intensity < 1.00E-002	100% β^-
Cd-109	3.81000E-02	4.00E+07	integral gamma intensity < 1.00E-002	100% β^+(m)
Cd-113	5.99000E-18	2.43E+23	pure alpha/beta	100% β^-
Cd-114	0.00000E+00[1]	1.89E+25	pure alpha/beta	100% 2 β^-
In-100	0.00000E+00[1]	5.9	pure alpha/beta	96.1% β^+, 1.95% β^+, p, 1.95% β^+,p (m)
...	...	...	...	...
Total contribution to hazard = 33.6%, which equals = 5.634330E-004 Bq/unit (13.6% of total activity)				

Data as such as those provided in the table above allows for providing a frame of reference of the criticality of radionuclides which cannot be measured by gamma-spectroscopy. ActiWiz 3 also provides such tables automatically for a set of fixed cooling periods (short term like 1 day up to long term like 30 years) in order to determine the maximum cool-down period beyond which the

[1] There is no contribution to the radiotoxicity as no clearance value is available for In-100 and Cd-114.

contribution of DTM nuclides become overwhelming. For elements like Indium such an analysis is critical as their contribution to the radiotoxicity can reach 99% after a few years of cooling.

2.2. Operational radiation protection

Maintenance periods at CERN are often directly preceded by machine development periods during which beam intensities can often vary. In operational Radiation Protection it is sometimes of interest to know up to which extent these deviations from normal operation can influence the nuclide inventory and related quantities like the radiotoxicity or clearance-level of samples. For this purpose the build-up and decay engine implemented in ActiWiz can treat also subsequent patterns of irradiation and cooling periods.

2.2.1. Complex irradiation & cooling patterns

As detailed in [2] ActiWiz implements a fully analytic solution to the Bateman differential equations which describe the build-up and subsequent decay of radioisotopes. Yet, this approach is not free from numerical pitfalls due to the limited floating point accuracy provided on today's CPU architecture. To overcome this shortcoming high-precision arithmetics with 512 significant bits are being utilized, which of course comes at the cost of performance.

The decay engine in ActiWiz is natively parallelized and will automatically utilize all cores using dynamic thread scheduling for maximum efficiency. Yet, the performance loss of switching from native double precision to arbitrary precision with 256 or 512 bits (~170 digits significant) results in a runtime penalty of about one order of magnitude or more. Yet, another possibility for optimization becomes apparent when analyzing the nature of the algorithm in more detail as some of the calculations involved in solving the Bateman equations can be considered repetitive if written in canonical form. This gives rise to the possibility of implementing a cache using a common "mailboxing technique". A more technical explanation is beyond the scope of this paper and can be found in Ref. [8].

In order to demonstrate the power of this concept the nuclide inventory of stainless steel has been determined considering one full month of SPS operation, modeling 55600 consecutive extraction and cool-down pairs (see Figure 4). In order to make the task somewhat more challenging the irradiation as well as the cool-down time have been randomly varied. While the overall period of 35 seconds had been retained, the irradiation time varied between 9 and 10 seconds, whereas the associated cool-down period was fixed at either 26 or 25 seconds. Such variations would be encountered if the duty and super-cycle of the accelerators were not integer numbers (e.g. 9,3 seconds) while the accuracy of the logging system's time stamps would be fixed to 1 second.

IOP Conf. Series: Journal of Physics: Conf. Series **1046** (2018) 012007 doi:10.1088/1742-6596/1046/1/012007

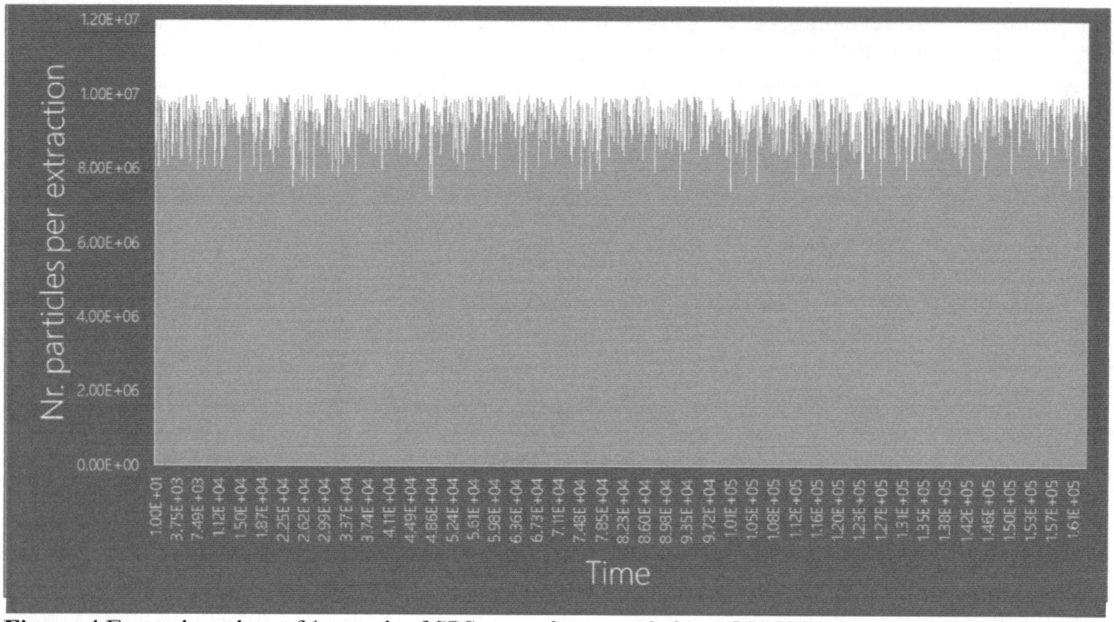

Figure 4 Exemplary data of 1 month of SPS operation, consisting of 55600 subsequent irradiation and cool-down pairs.

Using a standard desktop PC (i7-2600 quad-core at 3.4 GHz) the calculation could be performed in 93 seconds. While this example is extreme and somewhat academic, it serves to illustrate the potential of the implemented solution.

2.2.2. A practical application of an irradiation pattern

During the preparation of a maintenance stop of the Large Hadron Collider in 2016 it was discussed to reduce the beam intensity by a factor of about ~1/750 with respect to the nominal collision rate during the last four days (Thursday – Sunday) prior to the stop (Monday). However, it could not be excluded that at least on one day the intensity would not be raised again to nominal levels. In this specific case the interest was to determine the effect on the radiotoxicity of the C_6F_6 coolant used in the LHCb RICH detector. For this purpose four different scenarios have been evaluated modeling different irradiation/cooling patterns depicted in Figure 5. In all scenarios 6 months of continuous operation have been assumed as a basis followed either by four days of nominal operation, four days of low intensity operation or 3 days of low intensity operation interspersed with one day of nominal intensity.

For the purpose of modeling such time patterns ActiWiz provides a simple scripting language which allows for using mathematical expressions and variables (e.g.: 3 * day + 2 * hours – 5 * minutes). Within about twenty minutes it was possible to model and run a full study of four different scenarios yielding the results shown in Table 4.

IOP Conf. Series: Journal of Physics: Conf. Series **1046** (2018) 012007 doi:10.1088/1742-6596/1046/1/012007

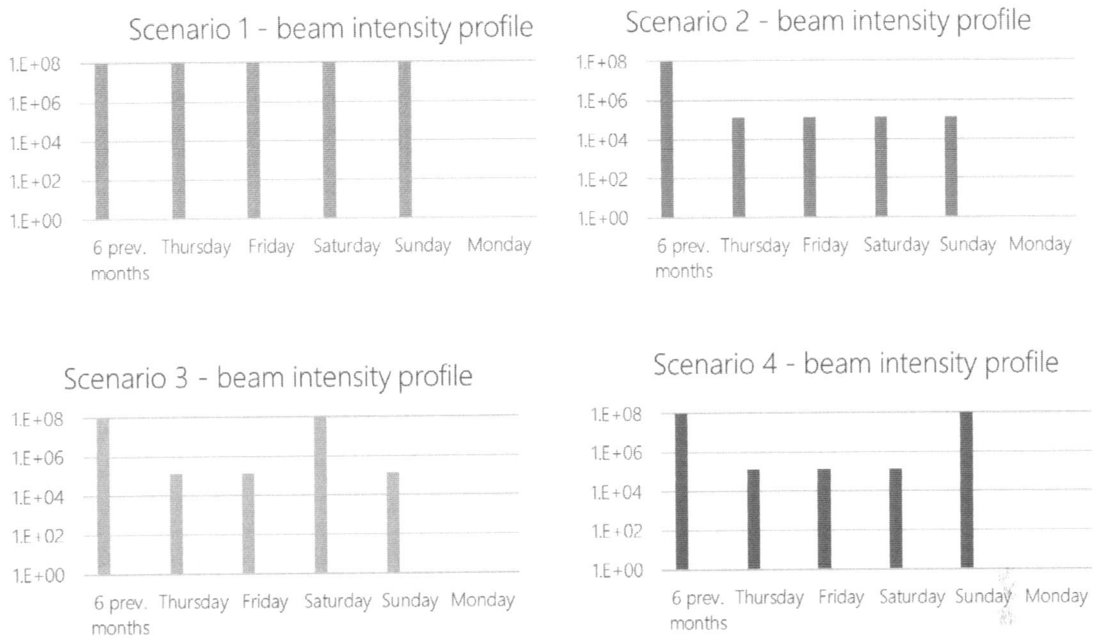

Figure 5 Four different irradiation / cooling scenarios prior to a maintenance stop of the LHC (Monday). In all scenarios 6 months of continuous operation have been assumed as a basis followed either by four days of nominal operation, four days of low intensity operation or 3 days of low intensity operation interspersed with one day of nominal intensity.

Table 4 Sum of the activity/clearance limit of the C_6F_6 coolant at LHCb's RICH detector as a function of the irradiation & cooling pattern prior to the machine stop.

Nominal intensity	Low intensity	Nominal intensity	Low intensity	Cooling	Activity / Swiss clearance limit
6 months	-	4 days	-	1 hour	0.0263
6 months	4 days	-	-	1 hour	0.00365
6 months	2 days	1 day	1 day	1 hour	0.0037
6 months	3 days	1 day	-	1 hour	0.0261

As can be seen only one day of low beam intensity is sufficient to decrease the radiotoxicity by a factor of about 7. Consequently, activation in the coolant is driven mainly by short-lived isotopes.

3. Summary and conclusions

This paper provides a brief overview of the ActiWiz 3 code which, in contrast to its predecessor, is not limited anymore to the assessment of nuclide inventories for CERN's high energy proton accelerators. Using particle fluence spectra for neutron, protons, charged pions as well as photons the software can assess nuclide inventories for compound materials exposed to arbitrary radiation fields within a few seconds. In addition the tool provides a large number of integrated analysis functionalities and out-of-the-box reports suitable for radiological characterization studies. In this paper features like the analysis of the radiotoxicity of thin Niobium coating as well as a radiation field dependent assessment of the

IOP Conf. Series: Journal of Physics: Conf. Series **1046** (2018) 012007 doi:10.1088/1742-6596/1046/1/012007

criticality of knowing the exact amount of Cobalt impurities in stainless steel have been illustrated. For this purpose the LEP super conducting cavities haven been used as an exemplary case. Furthermore, features like the automatic evaluation of the importance of difficult to measure nuclides have been briefly discussed before concluding with an exemplary application in operational RP for the preparation of maintenance works. As such the code complements the use of general purpose Monte Carlo codes like FLUKA as far as the rapid calculation and analysis of nuclide inventories is concerned.

References

[1] Vincke H and Theis C 2014 ActiWiz – optimizing your nuclide inventory at proton accelerators with a computer code *Proceedings of the ICRS12 conference, Nara, Japan, Progress in Nuclear Science and Technology* **4** 228-232

[2] Theis C and Vincke H 2015 The use of ActiWiz in operational radiation protection *Proceedings of the SATIF-12 conference at Fermilab 2014, Nuclear Science Report NEA/NSC/R* **3**

[3] Fassò A, Ferrari A, Ranft J and Sala P R 2005 FLUKA: a multi-particle transport code, *CERN Yellow Report, INFN/TC_05/11, SLAC-R-773*

[4] Battistoni G, Muraro S, Sala P R, Cerutti F,Ferrari A, Roesler S, Fassò A and Ranft J 2007 The FLUKA code: Description and benchmarking *Proceedings of the Hadronic Shower Simulation Workshop 2006, Fermilab 6--8 September 2006, M. Albrow, R. Raja eds., AIP Conf. Proc.* **896** 31-49

[5] Koning A, Forrest R, Kellett M, Mills R, Henrkisson R and Rugama Y 2009 The JEFF-3.1.1. nuclear data library *JEFF Report* **22**

[6] Vincke H and Theis C 2017 Mass production of nuclide production data used in the ActiWiz3 code *CERN Technical Note* **CERN-RP-2016-220-REPORTS-TN, EDMS 1761230**

[7] Theis C and Vincke H 2016 The LEP superconducting cavities – a practical example of how to use ActiWiz 3 for radiological characterization *CERN Technical Note* **CERN-RP-2016-170-REPORTS-TN, EDMS 1719831**

[8] Theis C and Vincke H 2016 ActiWiz 3 – an overview of the concepts, architecture & new features *CERN Technical Note* **CERN-RP-2016-117-REPORTS-TN, EDMS 1706010**

IOP Conf. Series: Journal of Physics: Conf. Series **1046** (2018) 012008 doi:10.1088/1742-6596/1046/1/012008

Induced activation studies for the LHC upgrade to High Luminosity LHC

C. Adorisio[1], S. Roesler[1]

[1] CERN, Radiation Protection Group, Geneva, Switzerland

cristina.adorisio@cern.ch

Abstract. The Large Hadron Collider (LHC) will be upgraded in 2019/2020 to increase its luminosity (rate of collisions) by a factor of five beyond its design value and the integrated luminosity by a factor ten, in order to maintain scientific progress and exploit its full capacity. The novel machine configuration, called High Luminosity LHC (HL-LHC), will increase consequently the level of activation of its components. The evaluation of the radiological impact of the HL-LHC operation in the Long Straight Sections of the Insertion Region 1 (ATLAS) and Insertion Region 5 (CMS) is presented. Using the Monte Carlo code FLUKA, ambient dose equivalent rate estimations have been performed on the basis of two announced operating scenarios and using the latest available machine layout. The HL-LHC project requires new technical infrastructure with caverns and 300 m long tunnels along the Insertion Regions 1 and 5. The new underground service galleries will be accessible during the operation of the accelerator machine. The radiological risk assessment for the Civil Engineering work foreseen to start excavating the new galleries in the next LHC Long Shutdown and the radiological impact of the machine operation will be discussed.

1. Introduction

1.1. Introduction to CERN

CERN, the European Organization for Nuclear Research, is an intergovernmental organization with over 20 Member States. Its seat is in Geneva but its premises are located on both sides of the French-Swiss border (http://cern.ch/fplinks/map.html).

CERN's mission is to enable international collaboration in the field of high-energy particle physics research and to this end it designs, builds and operates particle accelerators and the associated experimental areas. At present more than 11 000 scientific users from research institutes all over the world are using CERN's installations for their experiments.

The accelerator complex at CERN is a succession of machines with increasingly higher energies. Each machine injects the beam into the next one, which takes over to bring the beam to an even higher energy, and so on. The flagship of this complex is the Large Hadron Collider (LHC) as presented in Figure 1.

Further information is available on the CERN website: http://cern.ch.

IOP Conf. Series: Journal of Physics: Conf. Series **1046** (2018) 012008 doi:10.1088/1742-6596/1046/1/012008

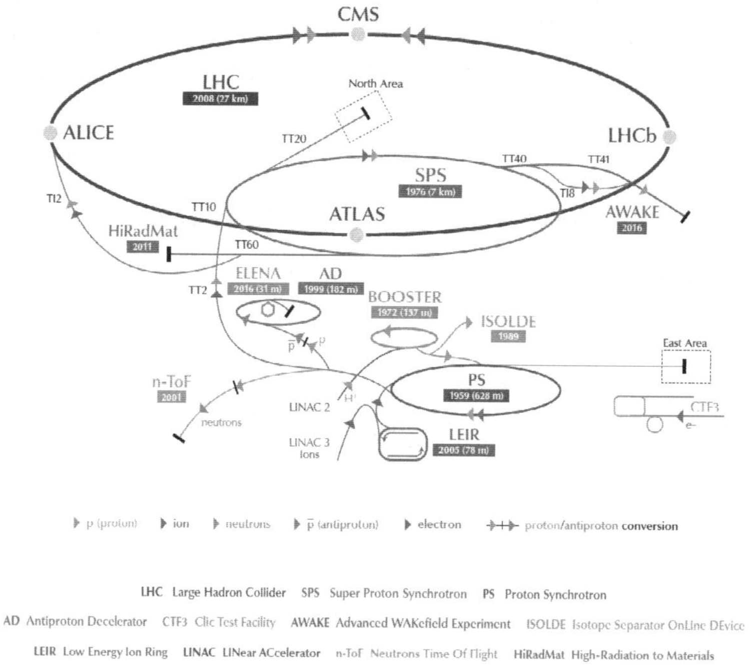

Figure 1. CERN accelerator complex.

1.2. Introduction to LHC and HL-LHC

The Large Hadron Collider (LHC) is the most recent accelerator constructed on the CERN site. The LHC machine accelerates and collides proton beams but also heavier ions up to lead. It is installed in a 27 km circumference tunnel, about 100 m underground.

High Luminosity LHC (HL-LHC) is a project aiming to upgrade the LHC collider during the Long Shutdowns (LS) 2 and 3, in order to maintain scientific progress and exploit its full capacity. By increasing its peak luminosity by a factor five over nominal value it will be able to reach a higher level of integrated luminosity, nearly ten times the initial LHC design target.

2. Radiation Protection studies using FLUKA Monte Carlo code

The radiological impact of the LHC proton-proton operation in the Long Straight Section (LSS) areas of the Insertion Region (IR) in Point 1 and 5 was evaluated in order to prepare the work for the modification of the LHC machine for its upgrade to HL-LHC. Using the Monte Carlo particle transport code FLUKA [1], [2], ambient dose equivalent rate estimations for the LSS regions have been performed on the basis of available operating scenario and using the latest available machine layout.

2.1. The p-p accelerator machine operating scenario

Proton-proton collisions in the LHC experiments in Point 1 (ATLAS) and in Point 5 (CMS) produce a secondary high energy radiation field that penetrates into the adjacent accelerator tunnels and causes severe activation of beam-line elements.

The simulations are performed considering the luminosity achieved up to 2016 and the target values announced by the LHC Operation team (LHC machine) and by the HL-LHC project (HL machine).

Table 1. Operating scenario of the LHC and HL-LHC machine.

	Year of (HL-)LHC Operation	Peak / levelled luminosity [cm^{-2}s^{-1}]		Integrated luminosity [fb^{-1}]	
past	≤2012	0.8E+34		30	
LS1					
past	2015	6.30E+33		4	
past	2016	1.40E+34		40	
	2017	1.60E+34		50	
	2018	1.60E+34		50	
LS2	(2 years)				
	2021	2.0E+34		60	
	2022	2.0E+34		60	
	2023	2.0E+34		60	~TOT 350 fb^{-1}
LS3	(3 years)	Nominal	Ultimate	Nominal	Ultimate
HL-LHC	2027	5.0E+34	7.5E+34	250	300
HL-LHC	2028	5.0E+34	7.5E+34	250	300
HL-LHC	2029	5.0E+34	7.5E+34	250	300
LS4	(1 year)				
HL-LHC	2031	5.0E+34	7.5E+34	250	300
HL-LHC	2032	5.0E+34	7.5E+34	250	300
HL-LHC	2033	5.0E+34	7.5E+34	250	300
LS5	(1 year)				
HL-LHC	2035	5.0E+34	7.5E+34	250	300
HL-LHC	2036	5.0E+34	7.5E+34	250	300
HL-LHC	2037	5.0E+34	7.5E+34	250	300
LS6				TOT ~2600 fb^{-1}	TOT ~3050 fb^{-1}

The operating scenarios as well as the integrated luminosity targets are summarized in *Table 1*.

2.2. *Simulation geometry*

LHC and HL-LHC geometry of the Insertion Region in Point 1 and in Point 5 has been reproduced in FLUKA [3], [4], all the machine elements as well as the tunnel are implemented in the simulation geometry.

The machine element layout and optics in the two points are the same, and from a radiological point of view the two Insertion Regions are equivalent. Figure 2 and Figure 3 show the horizontal cuts of the LHC and HL-LHC simulation geometry used, respectively.

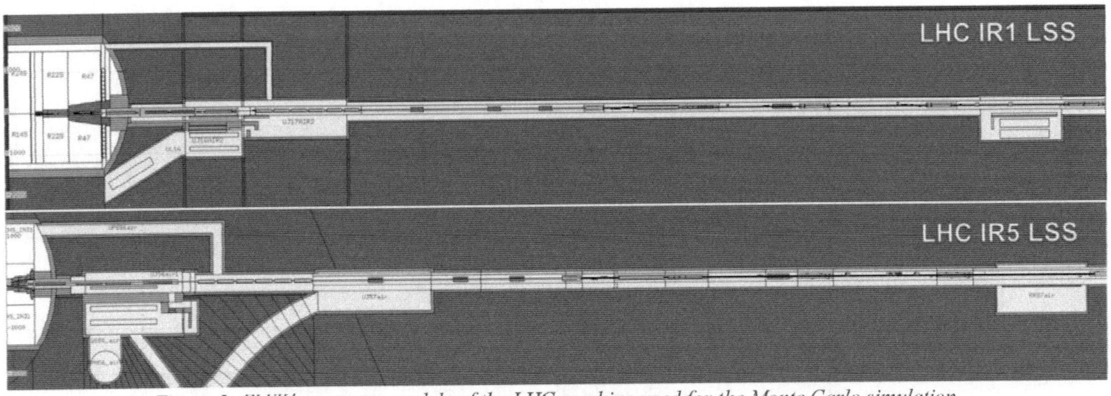

Figure 2. FLUKA geometry models of the LHC machine used for the Monte Carlo simulation.

IOP Conf. Series: Journal of Physics: Conf. Series **1046** (2018) 012008 doi:10.1088/1742-6596/1046/1/012008

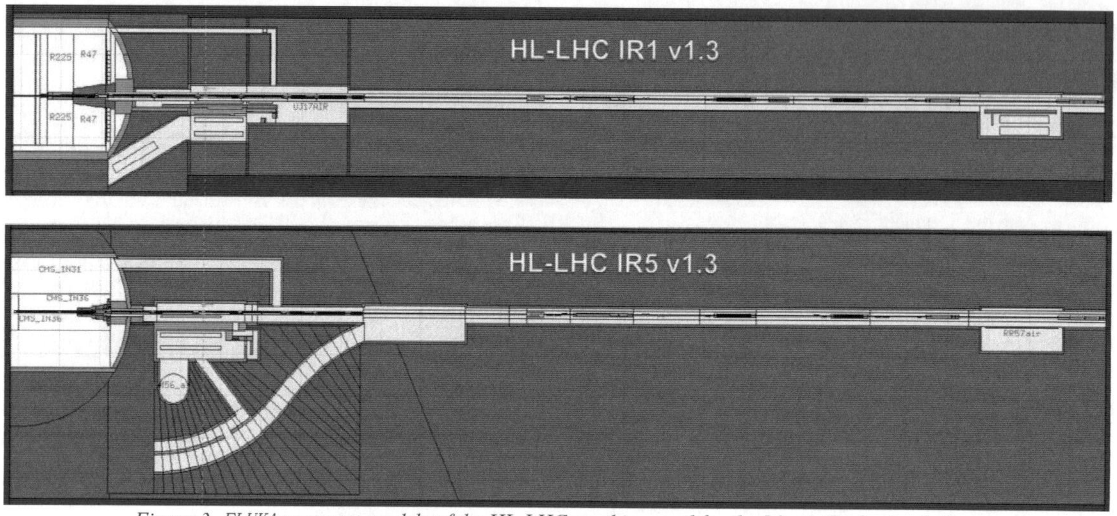

Figure 3. FLUKA geometry models of the HL-LHC machine used for the Monte Carlo simulation.

2.3. LHC machine simulation

Ambient dose equivalent rate maps have been calculated from around 18 m distance from the Interaction Point (IP) around the two main LHC experiment in Point 1 and 5 (ATLAS and CMS respectively) up to around 260 m distance, and for different cooling times in the LS2 and LS3 (as in Table 1). The simulations were done for 7+7 TeV pp-collisions and using DPMJET-III [5] as the event generator; for the inelastic proton-proton cross section the value of 85 mb is used, on the basis of the extrapolation from [6].

In order to benchmark the simulation results, during LHC machine operation, ambient dose equivalent rate maps have been calculated on the basis of the actual machine operating scenario and for typical cooling times during the Technical Stop of the machine.

For example, Figure 4 and Figure 5 show the comparison between the simulation results and the measurement taken in the LHC tunnel during the first Technical Stop in 2016. In the upper part of the plots the sequence of the machine elements is depicted.

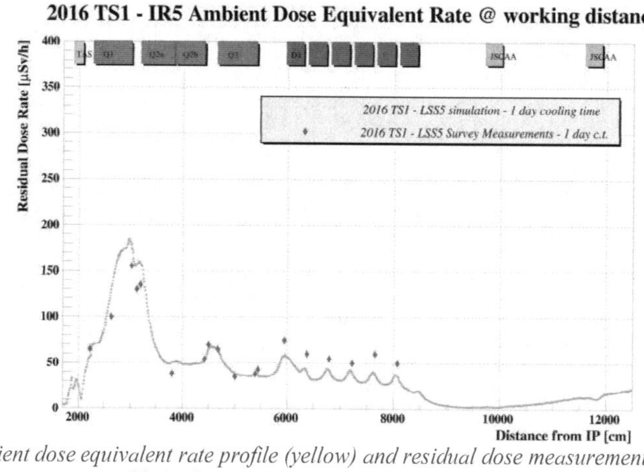

Figure 4. Calculate ambient dose equivalent rate profile (yellow) and residual dose measurements taken during LHC-TS1 (in blue) along the LHC machine elements in IR5.

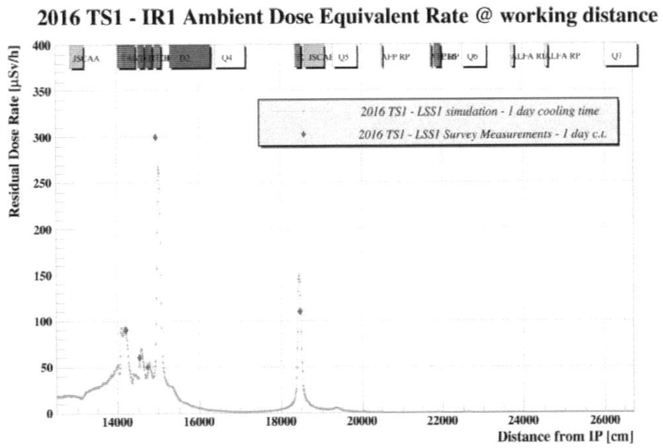

Figure 5. Calculate ambient dose equivalent rate profile (yellow) and residual dose measurements taken during LHC-TS1 (in blue) along the LHC machine elements in IR1.

The simulation results are in a very good agreement with the measurements.

Figure 6 shows the ambient dose equivalent rate map after one month of cooling time during LS3 in IR1. Figure 7 and Figure 8 show the residual dose rate profiles taken at a working distance (about 40 cm) from the machine elements in the tunnel aisle, at different cooling times (1 week, 4 weeks, 4 months and 1 year) during LS2 and LS3 respectively. The higher residual dose rates are around the elements closest to the IP (element on the left in plots), with the highest levels in the aisle corresponding to the connections between elements, where the shielding effect due to self-absorption of the element is less effective, and around the collimators (element in purple in the upper part of the profile plots, where the sequence of the machine elements is depicted).

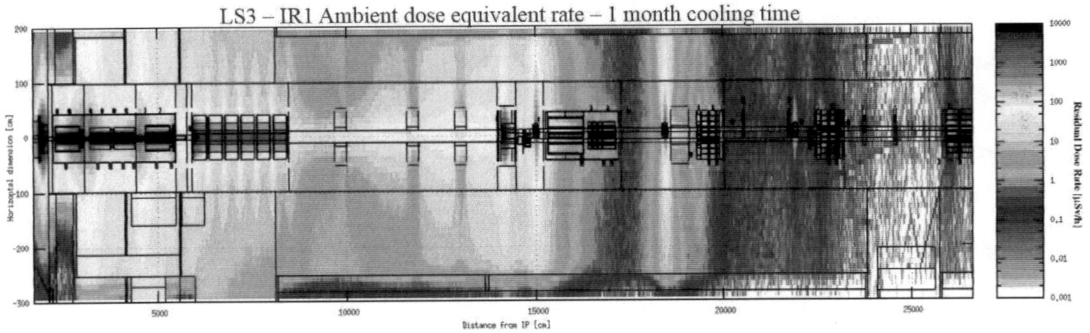

Figure 6. Ambient dose equivalent rate map in the Insertion Region 1 after one month cooling time during LHC machine Long Shutdown 3. The residual dose in the vertical dimension is averaged on 30 cm around the beam pipe.

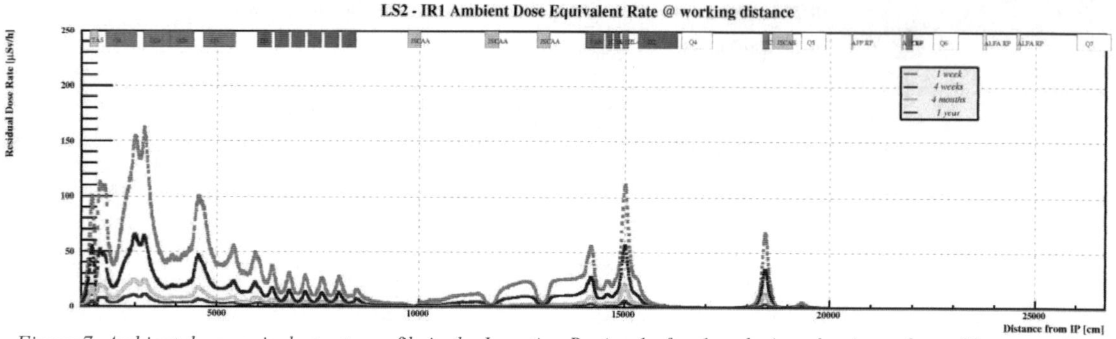

Figure 7. Ambient dose equivalent rate profile in the Insertion Region 1 after 1week, 4 weeks, 4 months and 1 year cooling time during LHC machine Long Shutdown 2. The residual dose in the vertical dimension is averaged on 30 cm.

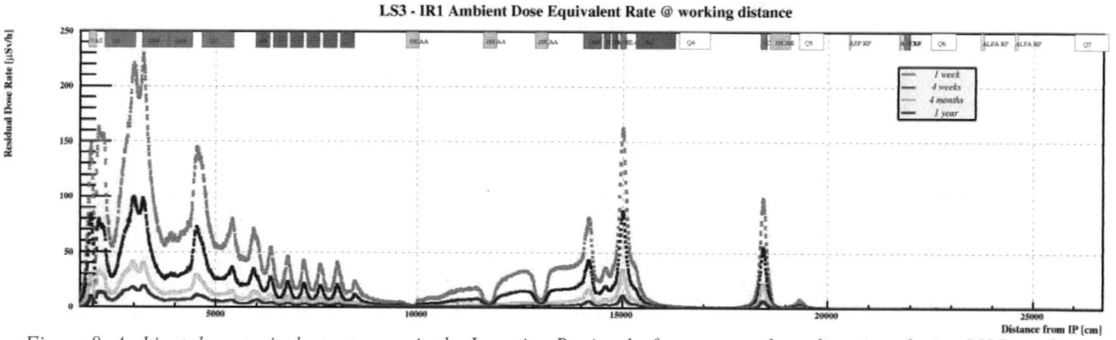

Figure 8. Ambient dose equivalent rate map in the Insertion Region 1 after one month cooling time during LHC machine Long Shutdown 3. The residual dose in the horizontal and vertical dimensions is averaged on 30 cm.

2.4. *HL-LHC machine simulation*

For the HL-LHC proton operation, the ultimate scenario is considered (7+7 TeV pp-collisions at the average luminosity value of 7.5×10^{34} $cm^{-2}s^{-1}$ with a 295 μrad half-angle crossing in the interaction point using DPMJET-III [5] as the event generator; for the inelastic proton-proton cross section the value of 85 mb is used, on the basis of the extrapolation from [6]).

Figure 9 shows the ambient dose equivalent rate maps after one month of cooling time during the first Long Shutdown during the HL-LHC era, so called LS4, in IR1 (upper plot) and in IR5 (bottom plot). The small difference in between the two ambient dose equivalent rate maps is due to the different crossing angle plane, which is vertical in IP1 and horizontal in IP5.

IOP Conf. Series: Journal of Physics: Conf. Series **1046** (2018) 012008 doi:10.1088/1742-6596/1046/1/012008

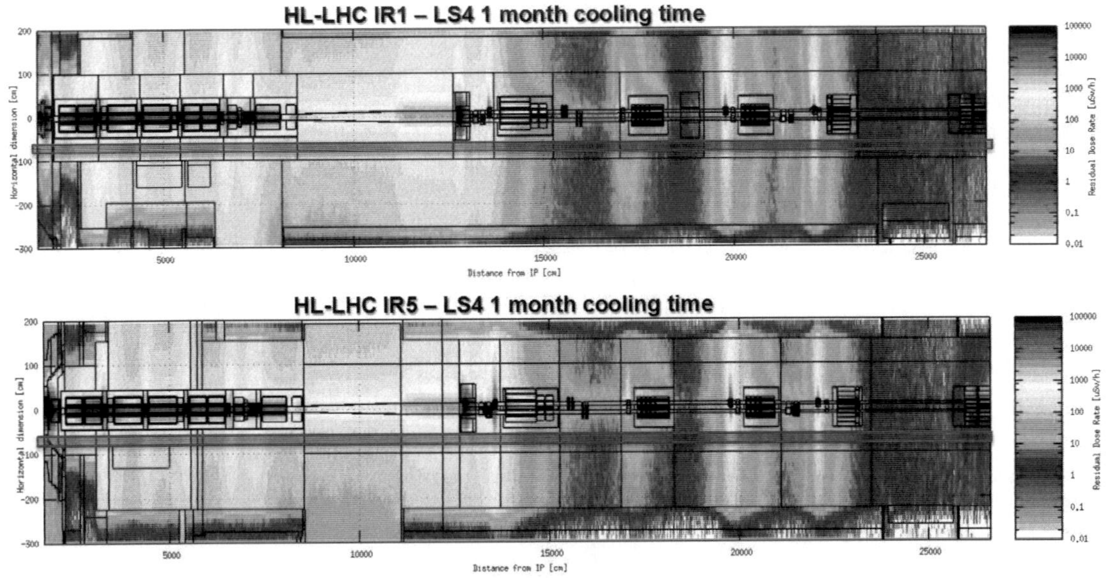

Figure 9. Ambient dose equivalent rate maps in the Insertion Region 1 (top plot) and in IR5 (bottom plot) after one month cooling time during HL-LHC machine Long Shutdown 4. The residual dose in the vertical dimension is averaged on 30 cm around the beam pipe.

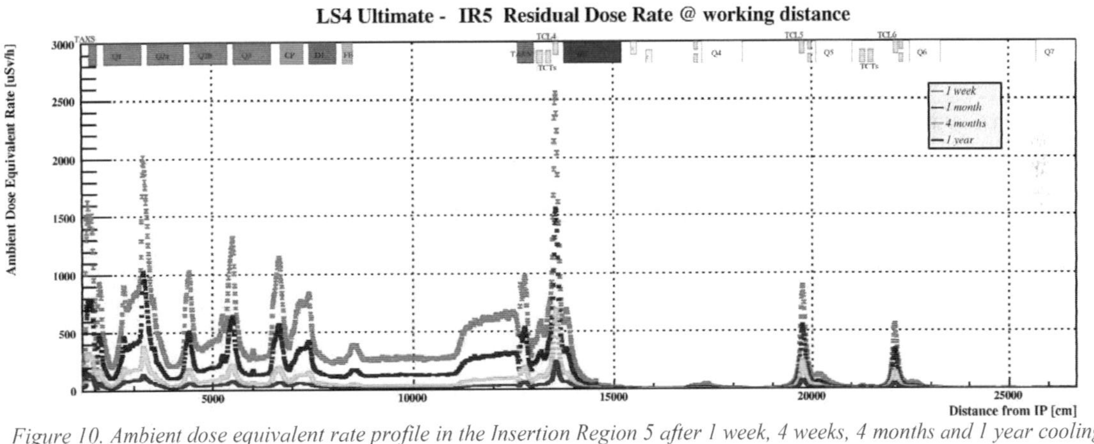

Figure 10. Ambient dose equivalent rate profile in the Insertion Region 5 after 1 week, 4 weeks, 4 months and 1 year cooling time during HL-LHC machine Long Shutdown 4. The residual dose in the horizontal and vertical dimensions are averaged on 30 cm.

Figure 10 shows the ambient dose equivalent rate profile in the aisle at a working distance from the machine elements at four different cooling times (1 week, 1 month, 4 months and 1 year), which are typical cooling times for maintenance intervention in the machine tunnel during the scheduled technical stops, end of the year technical stops and long shutdowns. As for the LHC machine, the higher residual dose rates are around the elements closest to the IP (element on the left in plots) where the highest levels in the aisle correspond to the connections between elements, where the shielding effect due to self-absorption of the element is less effective, and around the collimators (element in purple in the upper part of the profile plots, where the sequence of the machine elements is depicted).

As reported in Table 1, for the time being, three long shutdown are schedule for the HL-LHC. The operating scenario is foreseen as three blocks of three operational years with a long shutdown in between. For the operational years, constant operating parameters are assumed.

Figure 11 compares the ambient dose equivalent rate after one month of cooling time during the first HL-LHC long shutdown (LS4) and the last (LS6). The ratios for all the considered cooling times are reported in Table 2; the ratio is almost 1 for the shortest cooling time and it increase with the longer cooling times due to the accumulation in the activated equipment of the induced radionuclides with longer half-lives.

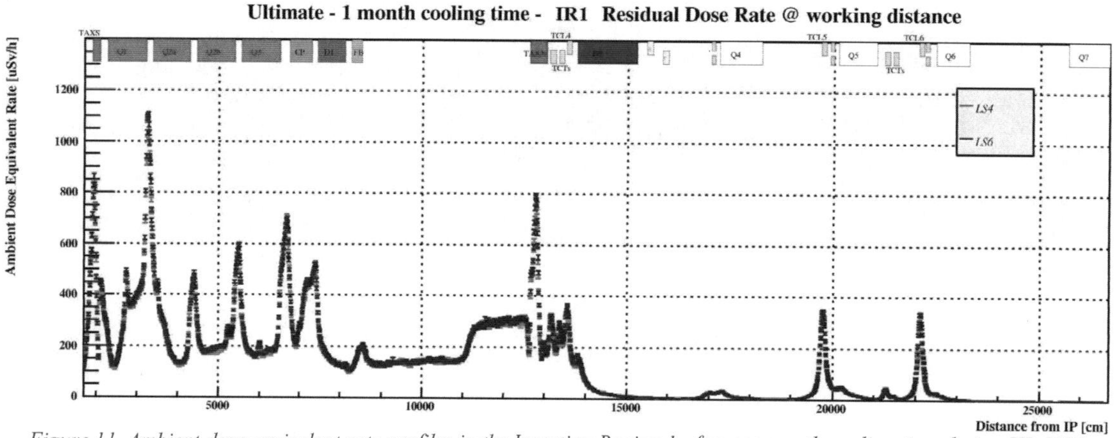

Figure 11. Ambient dose equivalent rate profiles in the Insertion Region 1 after one month cooling time during HL-LHC machine LS4 and LS6. The residual dose in the horizontal and vertical dimensions are averaged on 30 cm.

Table 2. Comparison residual dose rate ratios between LS4 and LS6 results for different cooling times.

Cooling time	LS6/LS4
1 hour	1.02
1 day	1.03
1 week	1.06
1 month	1.09
4 months	1.18
1 year	1.29

Figure 12 compares the ambient dose equivalent rate after one month of cooling time during the first HL-LHC long shutdown (LS4) between the nominal and the ultimate case (as reported in Table 1). The ratios for all the considered cooling times are reported in

Table 3; the ratio is higher for the shortest cooling time due to the higher production in the activated equipment of the induced radionuclides with shorter half-lives, and it decreases with the longer cooling times.

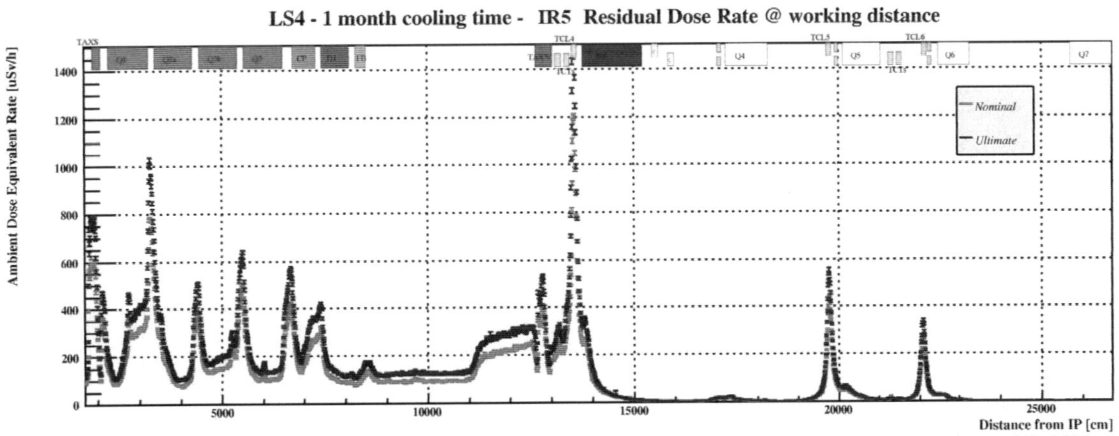

Figure 12. Ambient dose equivalent rate profiles in the Insertion Region 5 after one month cooling time during HL-LHC machine LS4 for the Nominal and Ultimate operating scenarios. The residual dose in the horizontal and vertical dimensions are averaged on 30 cm.

Table 3. Comparison residual dose rate ratios between Nominal and Ultimate results for different cooling times.

Cooling time	Ultimate/ Nominal
1 hour	1.50
1 day	1.46
1 week	1.35
1 month	1.30
4 months	1.24
1 year	1.23

2.5. *HL-LHC new underground galleries*

In order to accommodate the new technical infrastructure needed for the upgrade of the LHC machine to HL-LHC, new caverns and 300 m long galleries along the Insertion Region of IP1 (ATLAS) and IP5 (CMS) are required. The new underground service galleries, showed in red in Figure 13, will be on a higher level with respect to the LHC tunnel.

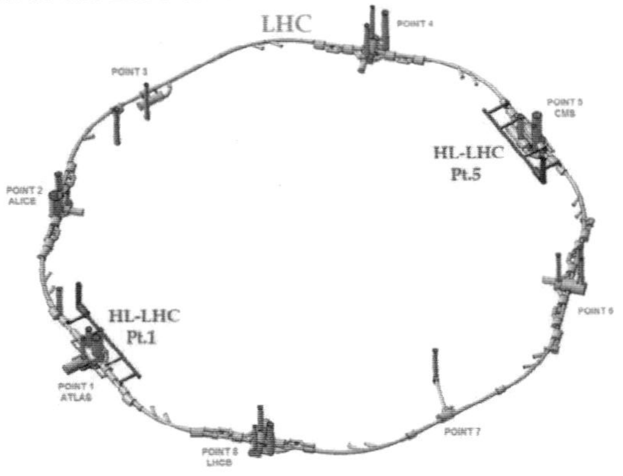

Figure 13. Schematic of HL-LHC with the new infrastructure galleries in red.

IOP Conf. Series: Journal of Physics: Conf. Series **1046** (2018) 012008 doi:10.1088/1742-6596/1046/1/012008

The new service galleries will be accessible during the operation of the accelerator machine. For this reason, the radiological impact of the machine operation has to be assessed.

Two different scenarios are kept into account in the study: an accidental scenario, represented by one full proton beam lost into a bulky object in front of the most exposing connection to the galleries, and the stray radiation from the normal operation of the machine. For both cases the ultimate beam intensity (2808 bunches and $2.2\text{x}10^{11}$ ppb) and luminosity ($7.5\text{x}10^{34}$ $cm^{-2}s^{-1}$) are considered. The requirements are set by the two considered scenarios. In the accidental case the effective dose delivered to personnel has not to exceed the legal annual limit for the class B radiation worker, i.e. 6 mSv. During normal operation of the accelerator machine, in the new service galleries the ambient equivalent dose has to be the lowest possible, and in any case, it has to be lower than 15 μSv/h, in order to classify the areas at the lowest Radiation Area level, i.e. Supervised Radiation Areas.

In Figure 14, the two scenarios and the corresponding operational parameters considered are summarized.

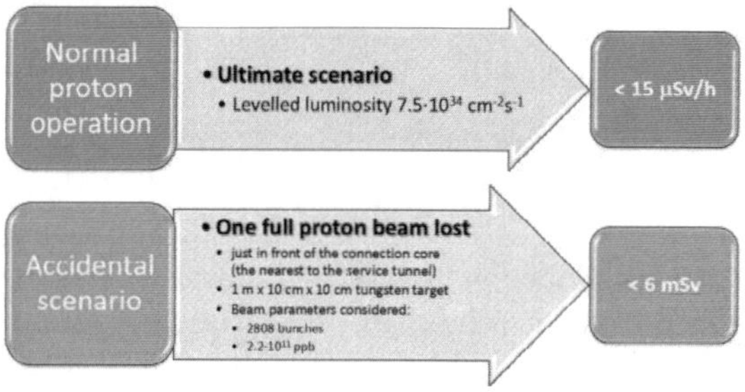

Figure 14. Operational parameters and scenarios considered.

In the FLUKA geometry model used for the previous LHC and HL-LHC studies (Ch. 2.3. and Ch. 2.4. in this proceeding), the new underground service galleries were not included at the time of this study. For this reason a standalone model of the service galleries was developed (Figure 15). In order to assess the impact of normal operation, the results from the accidental scenario study are used combined with the results from the study on the HL-LHC IR1.

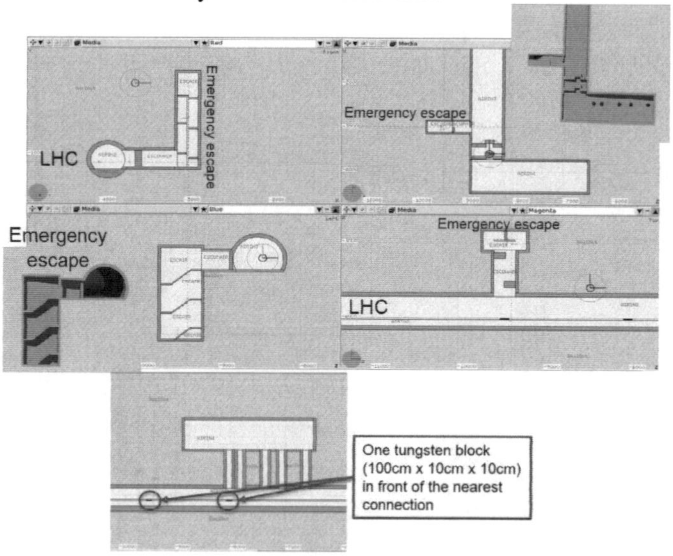

Figure 15. Different view of the new underground service galleries FLUKA model.

IOP Conf. Series: Journal of Physics: Conf. Series **1046** (2018) 012008 doi:10.1088/1742-6596/1046/1/012008

Figure 16 shows on the left the effective dose map and on the right effective dose profile in the area indicated by the red square, due to the full loss of a full proton beam on a bulky object in front a connection to the new galleries. The implemented shielding provide an attenuation of a factor of about 10^{-8} from the LHC tunnel to the closest accessible area in the new galleries (indicated by the blue line in profile plot, on the right of Figure 16), where the effective dose level drops to a delivered dose lower than 0.1 mSv.

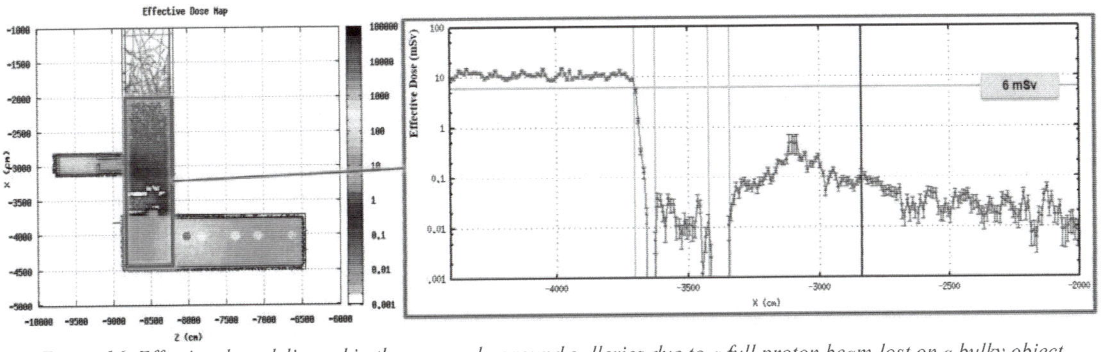

Figure 16. Effective dose delivered in the new underground galleries due to a full proton beam lost on a bulky object.

The ambient dose equivalent rate in the tunnel during the accelerator machine operation at ultimate intensity is shown in Figure 17, top part.

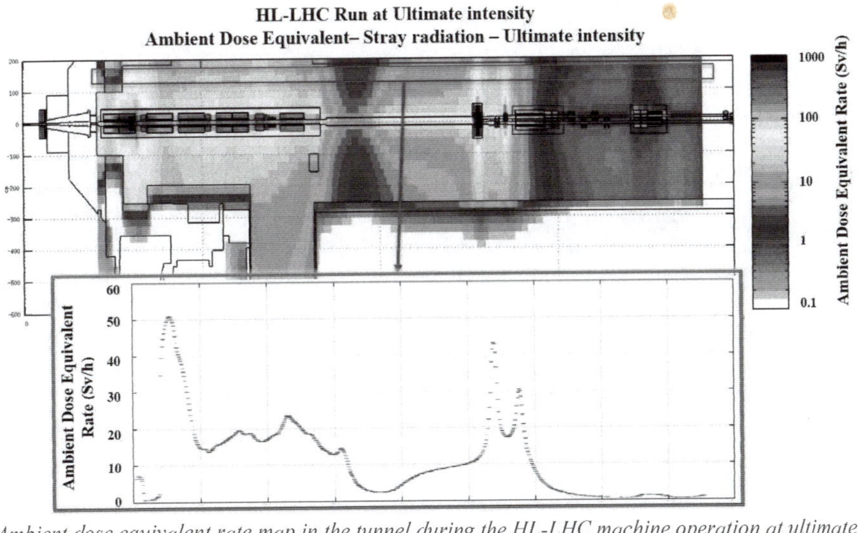

Figure 17. Ambient dose equivalent rate map in the tunnel during the HL-LHC machine operation at ultimate intensity, on top, and corresponding ambient dose equivalent rate profile next to the tunnel wall, on the bottom.

As can be seen in the profile plot on the bottom part of Figure 17, the maximum dose rate next to the tunnel wall is 50 Sv/h, which applying the attenuation factor of 10^{-8}, lead to a value of 0.5 μSv/h ambient dose equivalent rate due to stray radiation in the new underground gallery closest accessible point.

3. Activation studies using ActiWiz3 Creator code

In order to assess the level of activation of the spoil when the new underground galleries will be excavated (LS2), the results obtained with the FLUKA Monte Carlo code (particle fluence spectra) were used as input for the ActiWiz3 Creator © code [7].

An activated material is defined as radioactive if the specific or total activity of any radionuclides of artificial origin exceed the corresponding exemption limit, i.e.:

$$\sum_{i=1}^{n} \frac{a_i}{LE_i} < 1$$

where a_i is the specific activity (Bq/kg) or the total activity (Bq) of the ith radionuclide of artificial origin in the material, LE_i is the respective CERN exemption limit for the radio-nuclide i in the material and n is the number of radionuclides present. The CERN exemption limits (LE) are adopted from the Swiss legislation [6]. By the end of the current year (2017), these limits will be updated and the new version of this limits (LL) is taken into account for the estimation of the spoil activation in LS2.

During last end of the year technical stop of the LHC machine (EYETS 2016-17), some concrete and rock samples were taken out from the LHC tunnel wall, in the place where the excavation will occur (Figure 18.).

Figure 18. Concrete sample taken from the LHC tunnel wall during EYETS 2016-17.

γ-spectrometry measurements were performed on the rock and concrete samples, after a cooling time of about 3 months. The activation of the samples was also evaluated using ActiWiz3 Creator code. The results are reported in Table 4, in the form of the sum of specific activity over exception limit.

Table 4. Activation level of concrete and soil samples as measured by the g-spectrometry and calculated using the ActiWiz3 Creator© code.

$\sum_{i} \frac{a_i}{LE_i}$	**EYETS 2016-17 (~3 months cooling time)**		
UPR	γ spectrometry measurements	ActiWiz3 Creator©	
		only γ	*full*
concrete	1.15E-01	9.70E-02	*4.96E-01*
soil	8.17E-03	5.30E-03	*1.65E-02*

In the last column of Table 4, the results of all the radionuclide is reported, including the ones that cannot be measured with the γ-spectrometry technique. The γ results, second and third column in Table 4, are in good agreement. The induced radionuclide list calculated is reported in Table 5 (concrete sample) and in Table 6 (soil sample) with the contributing percentages to the sum; the most contributing radionuclide in both cases is Ca-45, which is a pure β-emitter, thus not measurable with the γ-spectrometry technique.

Table 5.

Concrete sample	
radionuclide	Contribution to $\Sigma_i \frac{a_i}{LE_i}$
Ca-45	80%
Na-22	7%
Zn-65	3%
Fe-55	3%
S-35	1%
Mn-54	1%

Table 6.

Soil sample	
radionuclide	Contribution to $\Sigma_i \frac{a_i}{LE_i}$
Ca-45	68%
Na-22	28%
S-35	2%

The activity over limit sum was evaluated for the spoils coming from the excavation of the new underground galleries where they are connected to the existing LHC tunnel (Figure 19).

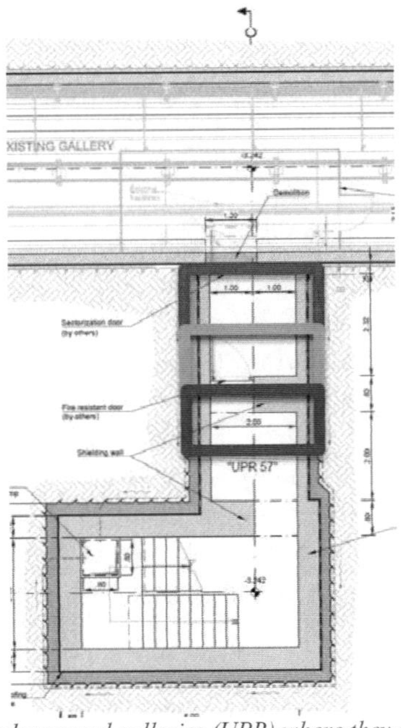

Figure 19. Snapshot of the new underground galleries (UPR) where they connect to the existing LHC tunnel.

The results of the calculation, for the first three meters of excavation spoils is reported in Table 7, for one and four months cooling times; the first meter of excavated spoils will be radioactive, the second and the third meters are below the legal limits in both the cases.

Table 7. Activation level of excavated spoils.

$\sum_i \frac{a_i}{LL_i}$	LS2	
UPR	ActiWiz 3 Creator©	
	1 month	4 months
1st meter	5.19E+00	4.15E+00
2nd meter	3.19E-01	2.90E-01
3rd meter	1.87E-02	1.70E-02

4. Summary

Fluka Monte Carlo code is extensively used to perform induced activation studies for the LHC accelerator machine and its High Luminosity upgrade. The simulations are performed using a very long and detailed geometry which reproduce the accelerator elements and the tunnel environment in great detail. The simulation results are continuously benchmarked with the radiation measurements performed in the LHC machine.

Actiwiz3 Creator© were used in order to predict the level of activation of excavation spoils and the radionuclide inventory in view of the excavation work in order to build the new underground galleries for the HL-LHC additional infrastructure and services.

References

[1] T.T. Böhlen, F. Cerutti, M.P.W. Chin, A. Fassò, A. Ferrari, P.G. Ortega, A. Mairani, P.R. Sala, G. Smirnov and V. Vlachoudis, "The FLUKA Code: Developments and Challenges for High Energy and Medical Applications," Nuclear Data Sheets 120, 211-214 (2014).

[2] A. Ferrari, P.R. Sala, A. Fasso`, and J. Ranft, ""FLUKA: a multi-particle transport code"," CERN-2005-10 (2005), INFN/TC_05/11, SLAC-R-773.

[3] V. Boccone et al., "Beam-machine Interaction at the CERN LHC," in *Nucl. Data Sheets 120 (2014), pp. 215-218, DOI: 10.1016/j.nds.2014.07.050.*

[4] A. Tsinganis, F. Cerutti, "Impact of Collision Debris in the HL-LHC ATLAS and CMS Insertions," in *in Proc. 8th Int. Particle Accelerator Conf. (IPAC'17), Copenhagen, Denmark, May 2017, paper TUPVA021, pp. 2093-2096, DOI:10.18429/JACoW-IPAC2017-TUPVA021.*

[5] S. Roesler, E. Enge and J. Ranft, *The Monte Carlo Event Generator DPMJET-III,* Vols. in Proceedings of the Monte Carlo 2000 Conference Lisbon, October 23-26, A. Kling, F. Barao, M. Nakagawa, L. Tavora and P. Vaz, Eds., Springer-Verlag Berlin, 1033-1038 (2001), 2000.

[6] Csörgő, T. et al., "Elastic Scattering and Total Cross-Section in p+p reactions measured by the LHC Experiment TOTEM at sqrt(s) = 7 TeV," in *XLI International Symposium on Multiparticle Dynamics*, 2012.

[7] C. Theis, H. Vincke, "ActiWiz 3 – an overview of the concepts, architecture & new features," in *CERN Technical Note CERN-RP-2016-117-REPORTS-TN, EDMS 1706010*, 2016.

[8] "Ordonnance sur la radioprotection (ORaP) du 22 juin 1994 (état le 1 janvier 2009)".

Saclay Compact Accelerator-driven Neutron Sources (SCANS)

A Marchix[1], A Letourneau[1], HN Tran[1], N Chauvin[1], A Menelle[2], F Ott[2] and J Schwindling[1]

[1] Irfu, CEA, Université Paris-Saclay, F-91191 Gif-sur-Yvette, France
[2] LLB, CEA, Université Paris-Saclay, F-91191 Gif-sur-Yvette, France

E-mail: anthony.marchix@cea.fr

Abstract. For next decade, the European neutron scattering community will face of important changes, as many facilities will close, strictly fission-based sources. This statement mainly concerns France with the planned closure of Orphee and ILL. At CEA-Saclay, the project SONATE has been launched in order to provide a high intensity neutron source in Saclay site, this project is based on Compact Accelerator-driven Neutron Sources technology coupled to high-intensity beams. The goal of SONATE is to develop a 50 kW target, aiming to produce at least a neutron yield of 10^{13} s^{-1} in pulse mode with a peak current of 100 mA. We have investigated in this document the best combinations of beam/target which would lead to this substantial neutron yields. Further investigations and tests have to be carry out, especially due to sparse data on thick target and such low-energy beams considered in this document. An intermediate step to the SONATE project is under test and development, called IPHI-NEUTRON, which would lead to provide a small-size neutron facility mainly devoted to neutron imagery for industry. This step is based on the existing 3 MeV proton beam, named IPHI. Best target candidates are Lithium and Beryllium, leading respectively to a neutron yield of about 2.10^{13} s^{-1} and 4.10^{12} s^{-1}.

1. Context

Neutron scattering is a very powerful method, widely used for studying materials properties in fundamental physics, materials science, nuclear energy, as well as for industries and societal challenges [1] [2]. Up to now, neutrons are not only produced from fission-based neutron sources (research nuclear reactors), but also from spallation sources. An overview of neutron scattering facilities in Europe is provided in Ref. [3], including their present status and future perspectives. This report brings to light that, for next decade, the European neutron scattering community will face of important changes, as many facilities will close, strictly fission-based sources. Unfortunately, no new project of nuclear reactor is, up to now, launched to replace them. Figure 1 illustrates this, showing a decreasing up to a factor of two on the beam time availability for the neutron scattering community in 2023, according to the baseline and degraded scenario, and 50% of decreasing in case of the enhanced scenario. French community is mainly affected with the closure of the Orphee reactor in Saclay site, which is planned end of 2019, as well as the ILL reactor in Grenoble with an uncertain future after 2023. An opportunity is open for Compact Accelerator-driven Neutron Sources (CANS), as their costs would be affordable for individual country (less than 200 Meuros), compared to the building of a dedicated reactor and they are more flexible in terms of management of targets and wastes, accelerator upgrading, safety constraints, etc. Those sources would be complementary to spallation and reactor sources in order to develop and test instrumentations, before going to those high-brilliance neutron sources, like ESS [4] for instance.

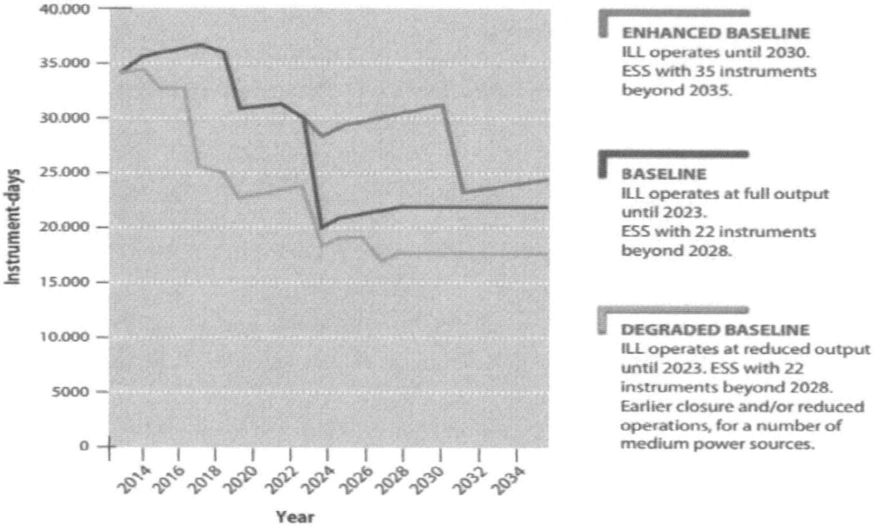

Figure 1: Scenarios for neutron provision in Europe (from Ref. [2]).

Compact Accelerator-driven Neutron Sources have knew a strong interest and a fast-growing, who led in 2009 to the foundation of the community UCANS (Union for Compact Accelerator-driven Neutron Sources) [5]. They are attractive alternative solutions to nuclear research reactors especially in terms of public acceptance as well as for nuclear non-proliferation. Many facilities around the world have already chosen this option but it concerns accelerators with few mA of continuous current. One can cite for instance the facility LENS in the United States [6] : 13 MeV proton beam, a peak current of 25 mA and a duty cycle of 1.8%, leading to an equivalent continuous current of 0.45 mA, i.e. a beam power of 6 kW.

Recent developments in high-intensity accelerators allow us to design high brilliance neutron sources, even with low-energy beams (few tens of MeV). This is for instance the case of the project IFMIF [7] for which the neutron production is induced by two 40 MeV deuteron beams striking on a liquid Lithium target. Note that each accelerator carries a beam current of 125 mA, leading to a total fast neutron yield of 10^{18} s^{-1}. At CEA/Irfu, 100 mA proton injector has been developed, called IPHI [8], as well as 125 mA deuteron injector, named SILHI [9]. The latter one is now part of the subproject IFMIF/LIPAc, the accelerator prototype of IFMIF. Both injectors may be operated in continuous or in pulse mode. High-intensity beam leads many challenges, especially due to strong space charge that can induce beam halo and consequently beam losses. Another issue concerns heat removal from the target. However, the mixing of concepts of high-intensity accelerator and compact moderator/reflector assembly permits not only to optimize the beam energy, and in consequence to reduce radiological constraints from beam-induced activation, but also to ensure an attractive neutron production.

In the following sections, we present potential neutron production reactions from low-energy beams, and experimental activities for our project. This involves not only the neutron yield, but also angular and energy distributions of emitted neutrons when available. We will also investigate regulatory constraints.

2. Project: objectives, constraints and roadmap

At CEA-Saclay, a project has been launched (SONATE), based on high-intensity accelerator technology and the concept of compact neutron source, using low-energy beam (less than 20 MeV). SONATE expects to provide to the neutron scattering community, as well as to local industries, a

small-scale neutron source with performances similar of a medium-size reactor (2 to 10 MW). Compared to spallation neutron sources, CANS are less efficient in producing neutrons, as the number of neutrons per incident particle is much less important. However, as neutrons produced by CANS are less energetic than those coming from spallation reactions, we may expect to convert them more efficiently into the required energy range and also to direct them more easily towards the required direction. This task would be facilitated as most applications require moderated or cold neutrons, thus a moderator, in association with a reflector, has to be optimized. Validated simulation tools are needed for the optimization of moderator/reflector geometry, taking into account neutronics properties of materials for particle transportation, including neutron diffraction within crystalline structures for thermal energy range, as well as in term of nuclear reaction cross sections.
First of all, our goal is to develop a target which is able to receive and dissipate 50 kW beam, in a compact layout, for the accelerator IPHI. This latter carries 3 MeV proton beam, 100 mA peak current, and is operated both in continuous and pulsed mode. Those developments are opportunity for validating our simulation tools, by measuring neutron flux according to different moderator shape and material [10]. This subproject, named IPHI-NEUTRON, is planned to operate in 2020, and expect to produce thermal neutron beams useful for neutron imagery applications, and small angle neutron scattering experiments. Even with the low neutron production related to 3 MeV protons, this project will be a demonstrator for the next step, SONATE, which could be in operation in 2025. Note that, we are currently limited to a power deposition density of about 0.5 kW/cm^2 on the target. This means that for a total power deposition of 50 kW, a target area of 100 cm^2 is needed. This is of importance for optimization studies associated to compact geometry of moderator/reflector assembly, as we need to concentrate as much as possible the neutron flux density within the moderator, as well as to reduce as much as possible neutron leakages.

Technique	Flux on sample	Reference spectrometers
Reflectivity	$0.8x10^7$ n/s/cm²	HERMES@LLB $1x10^7$ n/s/cm² POLREF@ISIS ~$1x10^7$ n/s/cm²
SANS	$0.7x10^6$ n/s/cm² (low Q) $2.2x10^6$ n/s/cm² (med Q) $6.7x10^6$ n/s/cm² (high Q)	PAXE@LLB (low Q) $0.7x10^6$ n/s/cm² SANS2D@ISIS $1x10^6$ n/s/cm²
Low resolution powder diffraction	$2x10^6$ n/s/cm²	G41@LLB $2x10^6$ n/s/cm²
Imaging (white beam)	$1.5x10^6$ n/s/cm² (for L/D = 240) $1.3x10^7$ n/s/cm² (for L/D = 80)	ICON@PSI $1x10^7$ n/s/cm² CONRAD@PSI $1x10^7$ n/s/cm² (for L/D = 240)
Imaging (time resolved)	$1x10^5$ n/s/cm² (for L/D = 500) dl/l = 1%	ANTARES@FRM2 $5x10^5$ n/s/cm²
Spin-Echo	$2x10^6$ n/s/cm²	MUSES@LLB $2x10^7$ n/s/cm² (at 5A°)
TOF	$6x10^7$ n/s/cm²	OSIRIS@ISIS $3x10^7$ n/cm²/s

Figure 2: SONATE's reference baseline in term of neutron fluxes and comparison to existing spectrometers.

Figure 2 shows the reference baseline of the project SONATE according to different measuring techniques. To fulfil these requirements, one needs a neutron yield roughly about 10^{15} s^{-1}, produced inside the target in continuous mode, and 10^{13} s^{-1} in pulsed mode by considering a peak current of 100 mA. Indeed, this leads to a neutron density within the moderator quite close to that one obtained from a medium-size reactor (10^{12} $cm^{-3}.s^{-1}$), taking into account a volume of 1 dm^3 for the compact moderator (instead of 10^6 cm^3 for the moderator volume of a reactor). Compared to the subproject IPHI-NEUTRON, SONATE's beam energy is higher in order to increase the neutron yield, but the total beam power is still limited to 50 kW. This affects the design of the moderator/reflector assembly, as the primary neutrons energy is higher as the beam energy increases. In consequence, a larger moderator is needed so as to be as efficient as the one used with 3 MeV protons (in term of neutrons moderation).

3. Neutron production induced by proton and deuteron beams

In this section, an overview of nuclear reactions leading to major neutron yields is presented. A discussion based on each step of the project, namely IPHI-NEUTRON and SONATE, provides the primary neutron characteristics (neutron yields, energy distributions, angular distributions) for these interesting reactions, when available. Only proton and deuteron beams were considered in the following parts based on the high-intensity beams available at CEA-Saclay. Moreover, only energy beams from reaction threshold up to 50 MeV concentrates our attention, as the goal is to take advantages from compact moderator layout. Note that, our discussion will concern thick targets in order to integrate beam reactions all along its range in the target, from incident energies down to the reaction thresholds.

3.1. Neutron production with IPHI-NEUTRON

Based on the accelerator IPHI that is carrying a 3 MeV proton beam, only few nuclear reactions may produce neutrons due to their low reaction thresholds. The reaction threshold strongly depends on the structure of the nucleus, and less bound neutron states (close to Fermi level) will lead to an enhanced microscopic cross section for the reaction (p,n). Indeed, those types of nuclei will emit more easily a neutron when a small excitation occurs. Concerning the neutron yield estimates, other parameters as the atom density of the target, the natural abundance of isotope of interest (if not available in enriched target), the effective range of protons in target, and the microscopic reaction cross section are also important. The effective range of protons means the range through the target for which the proton energy is upper than the reaction threshold, and in consequence leads to a neutron production. This proton range has been evaluated by using SRIM-2008 [11], and has to be as large as possible.

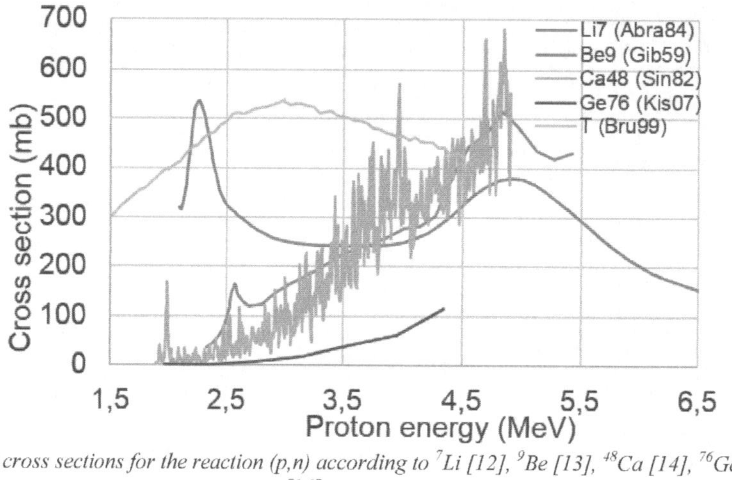

Figure 3: Experimental cross sections for the reaction (p,n) according to ^{7}Li [12], ^{9}Be [13], ^{48}Ca [14], ^{76}Ge [15] and Tritium [16] targets.

In Figure 3 are shown the experimental cross sections associated to the reaction (p,n), from their thresholds to few MeV, for ^{7}Li, ^{9}Be, ^{48}Ca, ^{76}Ge and Tritium targets. Usually, only light mass targets are considered for such beam energy, according to their low reaction thresholds and high cross sections such as Tritium, ^{7}Li and ^{9}Be. However, one may observe that higher mass isotopes, which have an excess number of neutrons, may also lead to a substantial neutron production. This is especially the case of ^{48}Ca which has a microscopic cross section for the (p,n) reaction similar to that one obtained for ^{9}Be. Table 1 provides necessary information to estimate roughly the neutron yields obtained for targets made of the specific isotopes considered in Figure 3, including their natural isotopic abundances. Table 1 also contains the energy threshold for the reaction (p,n). Note that Tritium is missing in this table, as that element has to be trapped into another material in order to form a thin layer on its surface [17][18]; Titanium or Scandium are mainly used for instance, especially for increasing heat deposition on such target, compared to enclosed tritium target. So, the neutron yield of a tritiated target is not so easy to estimate as for the ones considered in Table 1, as it depends on the Tritium concentration and layer thickness which is temperature-dependent. Nevertheless, commercial neutron generators based on the reaction $T(d,n)^4He$ are limited to a neutron yield of about 10^{13} s^{-1}, even if the cross section of this reaction exceeds largely those presented in Figure 3, with a maximum close to 5 barn for 100 keV deuteron beam. So, we may expect a neutron yield less important with 3 MeV protons striking on tritiated target than with 100 keV deuteron beam as its cross section is ten times less than the reaction $T(d,n)^4He$.

Target	^{7}Li	^{9}Be	^{48}Ca	^{76}Ge
Natural isotopic abundance (%)	92.5	100	0.2	7.6
Threshold (in MeV)	1.88	2.06	0.51	1.73
Atom density (cm^{-3})	4.59E22	1.24E23	1.93E22	4.22E22
Cross section (p,n) at 3 MeV (mb)	250	150	100	15
Effective range (µm)	182	46	165	36
Rough estimated neutron ratio (n/incident)	2.09E-04	4.27E-05	1.59E-05	1.14E-06

Table 1: Neutron ratio and associated important parameters for specific isotopes, considering 3 MeV incident protons and a thickness corresponding to the effective range.

First of all, we may observe from Table 1 that a Lithium target provides the highest neutron yield compared to other isotopes. This is due to its larger microscopic cross section for the reaction (p,n), as well as its longer effective range. For this element, we consider that the microscopic cross section at 3 MeV is constant all along the effective range, leading to an underestimation of the neutron yield since the cross section presents a higher value at 2 MeV compared to 3 MeV. On the other hand, according to the microscopic cross section behaviour for ^{9}Be, ^{48}Ca and ^{76}Ge, presented in Figure 3, the neutron yield has been estimated assuming a linear interpolation of the cross section along the effective range, starting from 0 up to the threshold. The neutron yield obtained with a target made of ^{9}Be isotopes is five times lower than for a Lithium target. Even if the cross sections for the reaction (p,n) are similar for ^{9}Be and ^{48}Ca, a target made of ^{48}Ca isotopes leads to a neutron yield which is three times less than a Beryllium target, despite the fact that the effective range in Calcium is four times greater than in Beryllium due to the compact crystalline structure of the Beryllium leading to a higher atom density. Note that an enrichment of 100% in ^{48}Ca isotopes is assumed for the Calcium target. Moreover, it seems difficult to operate a pure Calcium target as this element quickly oxidizes in air, becoming a powder. Nevertheless, for IPHI-NEUTRON, the neutron yield has to be maximized and in the following, only Lithium and Beryllium targets focus our attention.

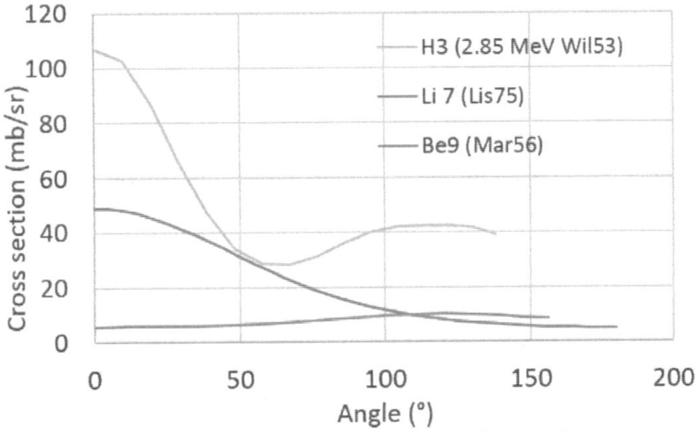

Figure 4: Experimental angular distributions for the reaction (p,n) for ^{7}Li [19], ^{9}Be [20] and Tritium [21] targets.

In the framework of the IPHI-NEUTRON sub-project, a target is under development that would be able to safely receive a total power deposition of 50 kW, i.e. a maximum beam current of 17 mA for 3

MeV proton beam. Consequently, this leads to a fast neutron yield of about 2.10^{13} s^{-1} for the Lithium target, and 4.10^{12} s^{-1} for the Beryllium target, the most favorable targets for the neutron production considering 3 MeV protons. Figure 4 shows the angular distributions of emitted neutrons associated to the Lithium and Beryllium targets for 3 MeV protons, and for the Tritium target for 2.85 MeV protons. In the case of Lithium and Tritium targets, we may observe that neutrons are preferentially emitted in forward direction, while for Beryllium target, the angular distribution is quite flat and favors neutrons in rearward (peak around 120°). This should affect the moderator efficiency and has to be taken into account in optimization studies, especially for defining the best position of the neutron guide. Concerning the energy distribution of the emitted neutrons, data are very sparse. Only one measurement is available for a thick Beryllium target, performed by Howard et al. [22] at two angles (0° and 40°).

Finally, we have also to discuss about the safety constraints of the potential targets (Lithium and Beryllium, and possible even a tritiated target), for 3 MeV protons. Both targets have their advantages and disadvantages. The Lithium target presents the advantage to be operated in a liquid phase, and in consequence to allow to operate at the maximum beam current of the accelerator IPHI, i.e. 100 mA. This option is attractive, as it permits to skip the heat removal issue while at the same time increasing the neutron yield. Such target has been developed by Soreq laboratory for instance that operates the loop LiLiT [23]. Nevertheless, one should notice that a huge production of ^{7}Be (about 1 MBq/s) occurs when operating a Lithium target. That element has a half-life of 53 days, and leads to constraints about waste management. Moreover, a liquid target provides more technical constraints associated to the loop, which may lead to potential neutron leakages and a less performance in term of neutron density within the moderator. The operating of a Beryllium target needs to worry about blistering effect that appears when beam is stopped inside the target [6]. To exclude this issue, the target thickness must be set to the effective range estimated in Table 1, i.e. 46 μm, in order to stop protons within a backing material which presents a better hydrogen diffusion coefficient. About tritiated target, the management of Tritium is an issue in case of failure of the target. Moreover, such target has a limited lifetime (about 1000 hours in operation), combining degraded performance during operation due to Tritium release from substrate material.

A Beryllium target seems to be the best choice, for 3 MeV protons, due to a substantial neutron production and less safety constraints. First experiment and test have been performed in 2016 with the accelerator IPHI. All the details can be found in [10][24], including a comparison between nuclear models and data on the primary neutrons emitted. No nuclear model is able to provide a reliable prediction by considering such low-energy beams. Else, a liquid Lithium presents huge advantages concerning the neutron production and also the heat removal issue, and further investigations have to be carry out especially about the optimization of the moderator/reflector assembly by including neutron leakages due to pipes of the liquid Lithium loop (performance reduction).

3.2. Neutron production for SONATE

A brief review is discussed in this section about potential nuclear reactions, induced by proton and deuteron beams, leading to a major neutron production for the project SONATE. We limit our investigations to beam energies less than 50 MeV, as well as to thick targets. To remind, the targeted neutron yield has to be roughly about 10^{15} s^{-1} in continuous mode, and 10^{13} s^{-1} in pulsed mode with a peak current of 100 mA, in order to fulfil the requirements of the project, whilst the beam power is limited to 50 kW. One has to notice that available data are very sparse and a comparison only based on such data is quite difficult to carry out. So, in this section, neutron yield estimates are obtained from Monte Carlo simulations, as only thick targets are considered. Calculations have been done using the Monte Carlo code MCNPX-2.6 [25], taking into account TENDL-2015 library [26] for proton-induced reactions below 200 MeV.

In Figure 5 are shown the neutron yields obtained for proton-induced reactions, considering 1 mA of beam current, for the following targets: Lithium, Beryllium, Carbon, Lead 208, Calcium 48, Germanium 76 and Tellurium 130. The three last isotopes have been considered because it refers neutron-rich isotopes. Those isotopes lead to the highest neutron production in the energy range 15 MeV to 30 MeV, according to TENDL-2015 library. Nevertheless, no data are available to confirm those estimates for thick targets. Note that Carbon target is not interesting for the neutron production, because of the reaction threshold of about 18 MeV for ^{12}C isotope; neutrons produced below 18 MeV only come from the reaction $^{13}C(p,n)^{13}N$, while ^{13}C represents only 1.1% of atoms within the target. Constrained to 50 kW beam power, this leads for instance to an equivalent continuous beam current of 3.33 mA for 15 MeV protons, 2.5 mA for 20 MeV protons and 1.66 mA for 30 MeV protons. Only the duty cycle has to be adjusted to fulfil these beam currents, while considering a peak current of 100 mA. Only neutron yields obtained with a Lithium and Calcium 48 target permit to reach the SONATE's requirement for a beam energy of 10 MeV (10^{13} s^{-1} in pulsed mode). For instance, a Beryllium or Lead target need a beam energy twice higher for the same performance. Keeping the beam energy as low as possible permits, first, to minimize the cost of the accelerator and the whole facility (safety and radiological constraints), as well as to limit the production of radiological contaminants (waste management and radioprotection).

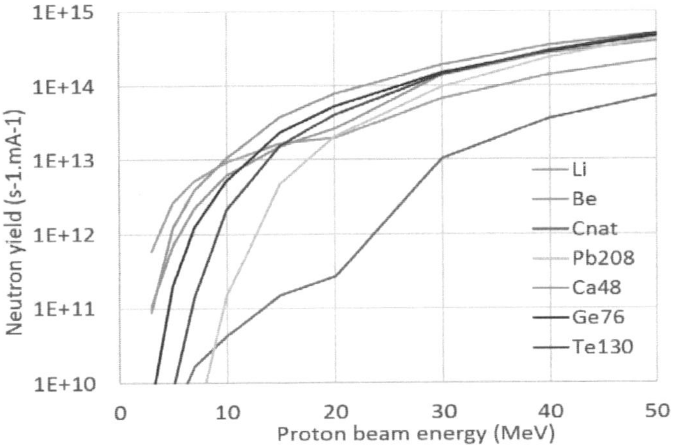

Figure 5: Neutron yield ($s^{-1}.mA^{-1}$) as a function of incident proton energy (MeV) obtained with TENDL2015 [26].

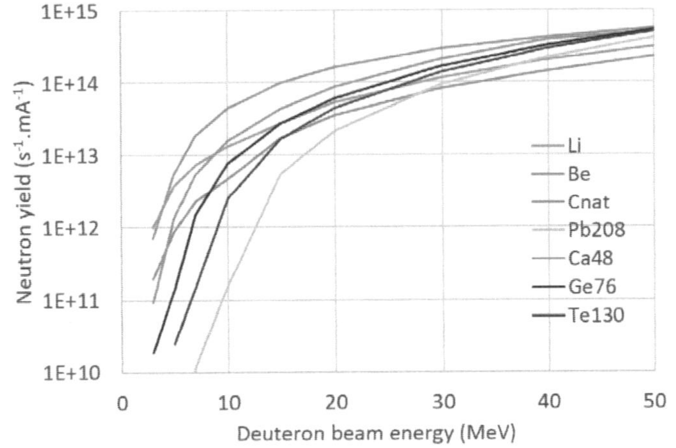

Figure 6: Neutron yield ($s^{-1}.mA^{-1}$) as a function of incident deuteron energy (MeV) obtained with TENDL2015 [26].

About deuteron-induced reactions, Figure 6 shows the neutron yield obtained for the same set of targets previously considered for proton beam. Deuteron beam energy has to be lower than that of protons in order to reach the same performance in term of neutron yield. Moreover, the Beryllium target induces the highest neutron yield for the whole energy range considered. With the Beryllium target, a deuteron energy of 7 MeV is enough to fulfil SONATE's requirement in term of pulsed neutron yield. This beam energy has to be roughly about 10 MeV for the Lithium and Calcium 48 targets, and around 15 MeV for Carbon. The lead target is only competitive with other studied targets for beam energies larger than 40 MeV. Compared to proton beam, neutrons produced by deuteron-induced reactions are mainly emitted in forward direction, due by part to the break-up of deuterons above 9 MeV, and also they are more energetic. Reference [27] provides the energy distribution and the angular distribution of neutrons produced by proton and deuteron beam, for a thick Beryllium target. More energetic neutrons will affect the geometry of the moderator by increasing its volume. This effect has to be considered for further optimization studies.

Depending on the production of radiological contaminants induced by beam-activation within the target, the best options for the project SONATE would be a proton beam striking on a target made of Calcium 48 isotopes or Lithium for an energy above 10 MeV, or a deuteron beam coupled to a Beryllium target with an energy above 7 MeV.

4. Conclusions and perspectives

Compact Accelerator-driven Neutron Sources are attractive alternatives to medium-size reactors so as to provide a local supply in neutrons for the European neutron scattering community. Those facilities would be affordable for individual country, as their cost depend on the beam energy which is limited as low as possible, first by using high-intensity beams, and also by considering pulse mode. At CEA/Irfu, proton and deuteron beams are able to be operated with high-intensity and in pulse mode. The project SONATE has been launched in order to provide a high intensity neutron source after the final shutdown of Orphee programmed end of 2019. This project is based on CANS technology coupled to high-intensity beams. The goal of SONATE is to develop a 50 kW target, aiming to produce a neutron yield of 10^{15} s^{-1} in continuous mode, and 10^{-13} s^{-1} in pulse mode with a peak current of 100 mA. We have investigated in this document the best combinations of beam/target which would lead to this substantial neutron yields. Best options for the project SONATE would be to consider a deuteron beam bombarding a Beryllium target with an energy larger than 7 MeV, or a proton beam coupled to a Lithium or Calcium 48 target for an energy larger than 10 MeV. Further investigations and tests have to be carry out, especially due to sparse data on thick target and such low-energy beams considered in this document.

An intermediate step to the SONATE project is under test and development, called IPHI-NEUTRON, which would lead to provide a small-size neutron facility mainly devoted to neutron imagery for industry. This step is based on the existing 3 MeV proton beam, named IPHI. Best target candidates are Lithium and Beryllium, leading respectively to a neutron yield of about 2.10^{13} s^{-1} and 4.10^{12} s^{-1}. A liquid Lithium target presents huge advantages concerning the neutron production and also the heat removal issue, even if this leads to a huge production of the radionuclide ^{7}Be (half-life of 53 days).

References

[1] IAEA 2005 *Technical report* IAEA-TECDOC-1439
[2] Nature Physics editorial March 2017 **13**
[3] ESFRI Physical Sciences and Engeneering Stategy Working group – Neutron Landscape Group 2016 *Neutron scattering facilities in Europe – Present status and future perspectives* (ESFRI Scripta **1**)
[4] Aberg M *et al* 2013 ESS Technical Design Report. European Spallation Source.
[5] Carpenter J M 2012 *Phys. Procedia* **26** 1
[6] Rinckel T *et al* 2012 *Phys. Procedia* **26** 161
[7] Knaster J *et al* 2015 *Nuc. Fusion* 55 086003
[8] Beauvais P Y 2004 *Proc. of EPAC* (Lucerne: Switzerland)
[9] Gobin R *et al* 2002 *Rev. Sci. Instrum.* **73** 922
[10] Letourneau A *et al* 2016 *Proc. ND2016 EPJ Web of Conferences* **146** 03018 (Bruges:Belgium)
[11] Ziegler J F *et al* 2010 *Nucl. Instr. And Meth. B* **268** 1818
[12] Abramovich S N *et al* 1984 *Yad. Konst.* **4** 17
[13] Gibbons J H *et al* 1959 *Phys. Rev.* **114** 571

[14] Singh G *et al* 1982 *Pramana J. Phys.* **19** 565
[15] Kiss G G *et al* 2007 *Phys. Rev. C* **76** 055807
[16] Brune C R *et al* 1999 *Phys. Rev. C* **60** 015801
[17] IAEA 2012 *Radiation Technology Reports Series* **1**
[18] Verbeke J M *et al* 2000 *App Rad. Iso.* **53** 801
[19] Liskien H *et al* 1975 *Atomic Data and Nuc. Data Tab.* **15** 57
[20] Marion J B 1956 *Phys. Rev.* **103** 713
[21] Willard H B *et al* 1953 *Phys. Rev.* **90** 865
[22] Howard W B *et al* 2001 *Nuc. Sci. And Eng.* **138** 145
[23] Halfon S *et al* 2011 *App. Rad. Iso.* **69** 1654
[24] Tran H N *et al* 2016 *to be published in Proc. Conf. SATIF-13* (Dresden: Germany)
[25] Pelowitz D B *et al* 2008 *User's manual version 2.6.0* Los Alamos National Laboratory Report LA-CP-07-1473
[26] Koning A J *et al* 2012 *Nuc Data Sheets* **113** 284
[27] Brede H J *et al* 1989 *NIM A* **274** 332

IOP Conf. Series: Journal of Physics: Conf. Series **1046** (2018) 012010 doi:10.1088/1742-6596/1046/1/012010

Study of neutron shielding collimators for curved beamlines at the European Spallation Source

V. Santoro [1,2], D. D. DiJulio[1,2], S. Ansell[1], N. Cherkashyna[1], G. Muhrer[1] and P. M. Bentley[1,3]

[1] European Spallation Source ERIC, 22100 Lund, Sweden
[2] Department of Physics, Lund University, 22100 Lund, Sweden
[3] Department of Physics and Astronomy, Uppsala University, 75105 Uppsala, Sweden

E-mail: valentina.santoro@esss.se

Abstract. The European Spallation Source is being constructed in Lund, Sweden and is planned to be the world's brightest pulsed spallation neutron source for cold and thermal neutron beams ($\leq$ 1 eV). The facility uses a 2 GeV proton beam to produce neutrons from a tungsten target. The neutrons are then moderated in a moderator assembly consisting of both liquid hydrogen and water compartments. Surrounding the moderator are 22 beamports, which view the moderator's outside surfaces. The beamports are connected to long neutron guides that transport the moderated neutrons to the sample position via reflections. As well as the desired moderated neutrons, fast neutrons coming directly from the target can find their way down the beamlines. These can create unwanted sources of background for the instruments. To mitigate such a kind of background, several instruments will use curved guides to lose direct line-of-sight (LoS) to the moderator and the target. In addition instruments can also use shielding collimators to reduce the amount of fast neutrons further traveling down the guide due to albedo reflections or streaming. Several different materials have been proposed for this purpose. We present the results of a study of different options for collimators and identify the optimal choices that balance cost, background and activation levels.

1. Introduction

The European Spallation Source (ESS) will be the most powerful neutron source in the world for neutron scattering studies of condensed matter. Neutrons will be produced by a 2 GeV proton beam impinging on a target made of tungsten. The neutrons will be eventually slowed down by the moderator placed next to the target. A reflector will surround the moderator to increase the neutron intensity. The tungsten target, moderator and reflector will be located inside the target monolith that will be shielded by a 3.5 m of steel which extends to a radius of 5.5 m from the moderator center. Within this monolith shield wall there will be beam extraction ports for each instrument. The monolith wall will shield the target and reduce the radiation escaping from it. Nevertheless, the neutron radiation dose coming out from the monolith is still substantial and must be shielded and contained in an additional common shielding structure called the bunker.

The ESS bunker will extend from the outer surface of the monolith located at 5.5 m from the target to 15 m for the short instrument (instruments several tens of meters) and to 28 m for long instruments (instruments up to $\sim$150 m in length). The current design of the ESS bunker is essentially a contained void with a 1.7 m thick roof about 3 m above the internal bunker floor and a 3.5 m thick wall, starting at 11.5 m for the short instrument sector or 24.5 m for the long instrument sector, as can be seen in figure 1.

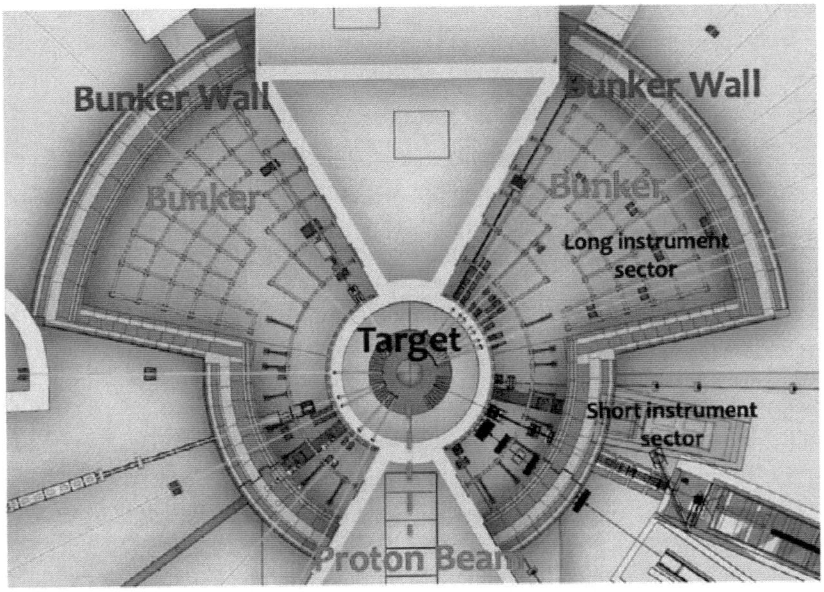

Figure 1: The engineering model of the ESS bunker.

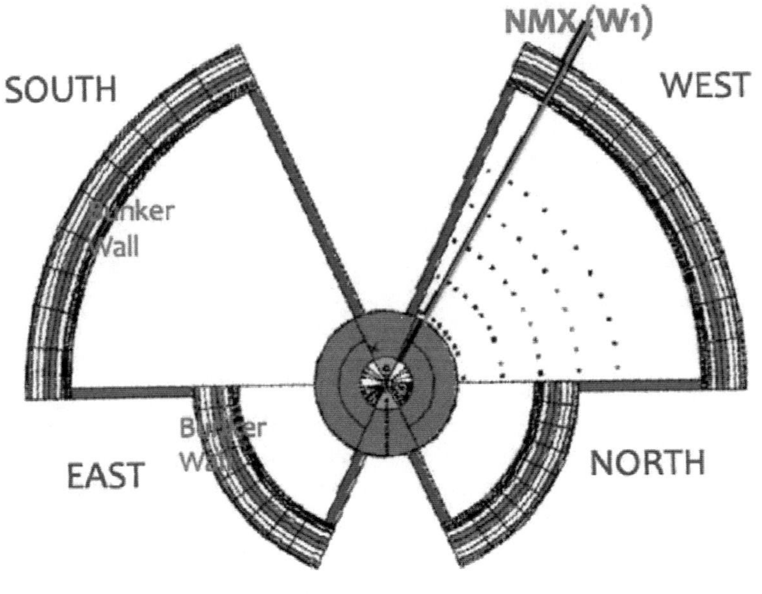

Figure 2: The MCNP6 model of the ESS bunker. All the sectors have been included, the NMX beamline is modeled on the first beam line on the west sector (W1).

The open area inside the bunker will be filled with components related to the instruments such as metallic guides, choppers, shutters or collimators. These components will receive a substantial radiation dose during beam on target and they will be activated. Rapid access to the equipment in this area is required for performing maintenance and repairs. For this reason, the optimisation of the activation level plays an important role in determining the serviceability and time-scales for accessing these components. In

IOP Conf. Series: Journal of Physics: Conf. Series **1046** (2018) 012010 doi:10.1088/1742-6596/1046/1/012010

Table 1: NMX beamline parameters as used in the simulation.

System	Description						
Beam properties at sample position	Wavelength band			Beam size		Divergence	
	1.74 Å (1.8-3.55 Å)			5mm x 5 mm (max)		± 0.2° (max)	
Neutron Guides	Start (m)	End (m)	Height (cm)	Width (cm)	Horizontal Geometry	Vertical Geometry	Coating
Inside target vessel	2.0	6.0	3.0 to 4.5	3.0	Straight	Lin. tapered	m=1 sides; m=2 top/bottom
Inside Bunker	6.0	31.5	4.5	3.0	Curved R1.2Km	Straight	m=1; m=2.2 only outer side,
Outside Bunker	31.5	154	4.5	3.0	Straight	Straight	m=1

this paper we focus on the neutron shielding collimators for curved beam lines. The choice of material for such a component is a compromise between the efficiency of reducing the neutron radiation dose, the cost, and the activation of the material.

2. The NMX instrument

For the collimator optimisation study we will use the NMX [1] instrument that is a time-of-flight (TOF) quasi-Laue diffractometer optimised for small samples and large unit cells. A table with the beamline parameters in its original design and used in this simulation can be found in table 1. The NMX instrument is basically composed of a straight guide in the monolith and a bender in the bunker. The curving of the guide permits the loss of line of sight to the moderator and target and the use of collimators allows the stripping out of the fast neutron component traveling down the guide.

To perform these studies we used the radiation transport code MCNP6 [2]. The MCNP6 model has been built using Combiayer [3] which is a C++ package that can write MCNP input files with a detailed geometrical description. A snapshot of the MCNP6 model for the ESS created with Comblayer is shown in figure 2. As can be seen from the picture the model contains several details like the exact geometry of the target wheel, the moderator and the monolith and is a very close representation of the current engineering model (figure 1).

The NMX beamline is located in the long instrument west sector and since it is the first beamline of this sector it is identified with the beamport W1. The beamline has been modelled in details with the guide in the monolith, the vacuum pipe, the vacuum window and the guide in the bunker, with the addition of two collimators. Figure 3 shows an overview of the beamline, while figure 4 shows the possible locations for the collimators in the bunker. As stated above, the primary objective of this study is to design the collimator blocks to minimise the fast neutron component that is travelling inside the guide. Reducing this radiation is not only advantageous for background reasons but can also considerably reduce the shielding cost further downstream of the collimator position.

3. Optimisation of the collimators blocks

To study what is the best configuration that allows a reduction of the fast neutron component we calculate the neutron energy spectrum immediately after the bunker wall (28 m after the moderator center) in the guide, for different collimators options and positions. The results of these calculations can be seen in figure 5, 6 and 7. The use of collimators, as can be seen from the spectra shown in figure 5, significantly reduces the high energy component but there is also a reduction in the keV energy region of the spectrum. Among all the possibilities studied, the best option is to use one collimator around 16 m where the guide started to lose significantly line of sight. Locating a collimator upper stream (12 m from

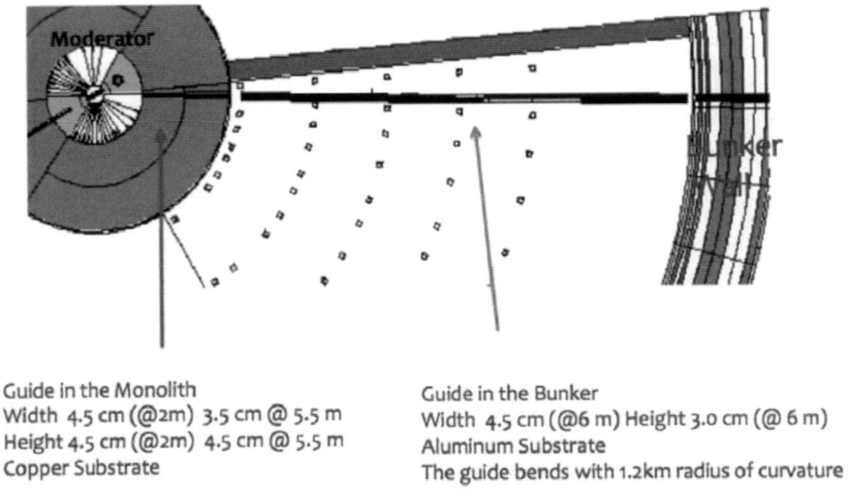

Figure 3: Overview of the NMX beamline in the bunker .

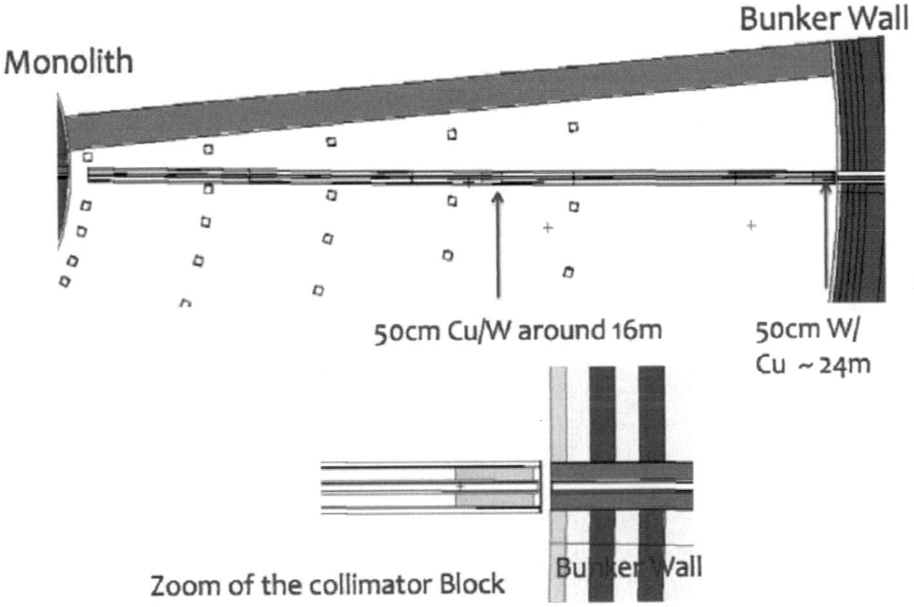

Figure 4: Collimator blocks in the bunkers. In the current picture one collimator is located at 16 m from the moderator and the other one is just before the bunker wall (24 m from the moderator).

the moderator) or downstream (20 m from the moderator) does not have the same effect. Adding another collimator just before the bunker wall has no significant effect (from figure 5 the difference between the spectrum for one collimator at 16 m (red dots) and the one with the addition of a second collimator at 24 m (blue dots) can not be seen) since the bunker insert is sufficient to remove neutrons that have not been stripped off by the first collimator. In all these studies, the length of the collimator was 50 cm and the material used was pure tungsten. An additional calculation to establish what is the effect of decreasing or increasing the length of the collimator is shown in figure 6. The results show that a 30 cm collimator provides the same performance as a 50 cm or 80 cm collimator. Thus increasing the length beyond 30 cm adds no performance.

Once we found the best position and the ideal length for the collimator, we compared different materials. The plot in figure 7 shows a comparison between pure copper, pure tungsten and mild steel (mild steel is a low carbon steel, its composition is shown in table 7). As can be seen in the figure, the performance of the three materials is the same.

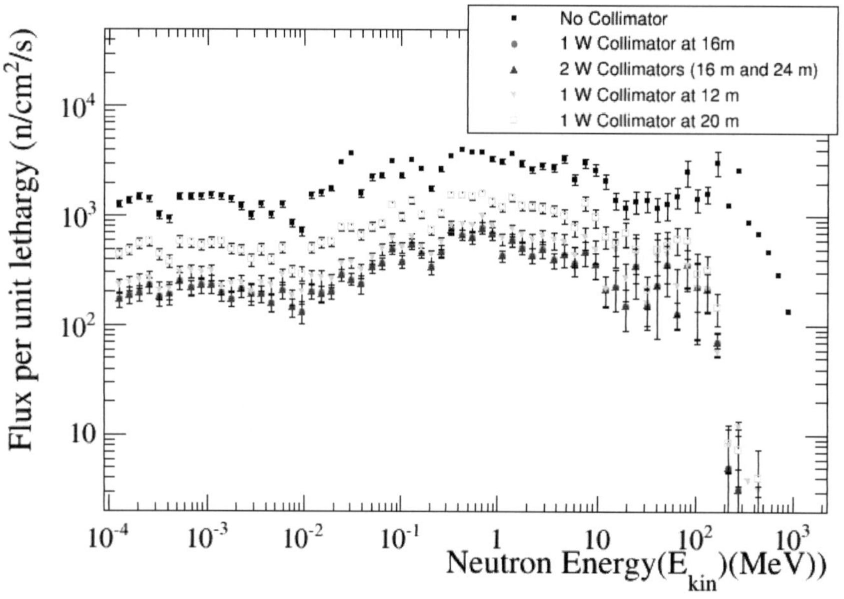

Figure 5: Fast neutron energy spectrum for different collimator configurations. The difference between the neutron energy spectrum from one tungsten collimator at 16 m (red dots) and the use of an additional collimator at 24 m (blue dots) could not be seen since they completely overlap. The reason is due to the fact that the bunker insert is sufficient to remove neutrons that have not stripped off by the first collimator.

4. Activation of the collimators block

We observed that the fast neutron reduction for the collimators are very similar for different materials. To find the best option however it is necessary to study their activation levels after the beam is off, since as explained before the collimators will be located in the ESS bunker area where maintenance and repairs of different components will be carried out. All the activation calculations have been performed using CINDER 1.05 [4]. For this calculation we used pure tungsten and pure copper, while for the steel we used a low carbon steel. The composition of the steel can be found in table 2. The activation levels

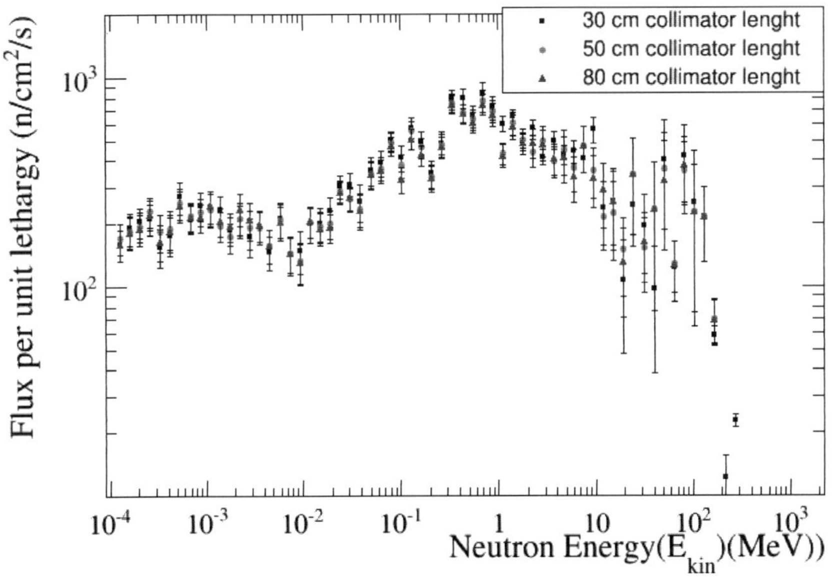

Figure 6: Fast neutron energy spectrum for different collimator lengths.

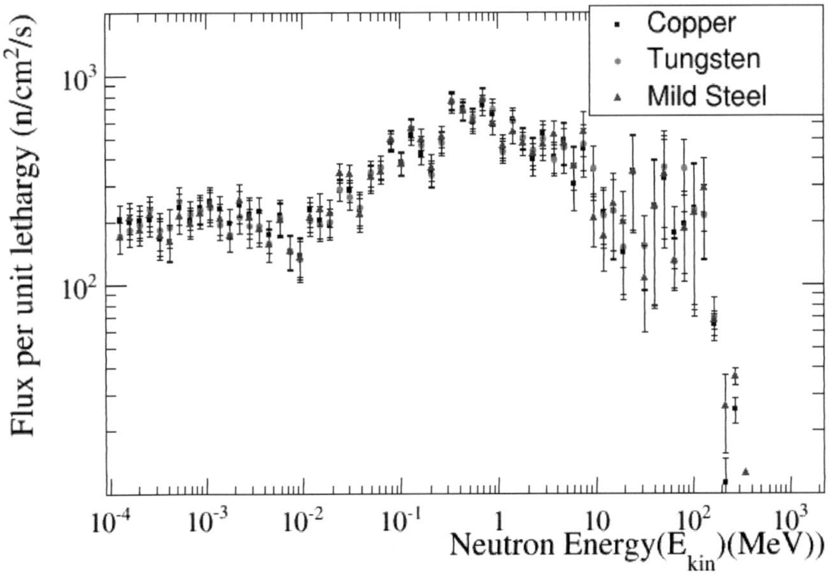

Figure 7: Fast neutron energy spectrum for different collimator materials. The tungsten and the copper used correspond to pure material while the steel composition is given in table 2.

were studied using the collimator located at 16 m and by dividing it into several different segments (see figure 8) since the first part of the collimator will receive a higher flux of neutrons and become more active.

The activation of the collimator block is computed after 10 years of irradiation with the proton beam

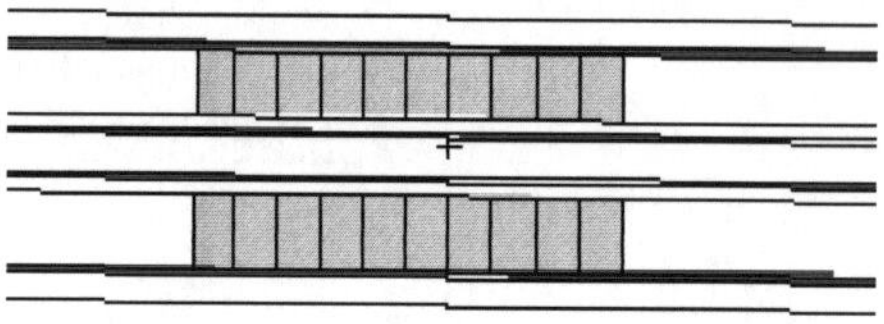

Figure 8: Side view of the collimator. The collimator has been divided into several different segments .

energy of 2 GeV at the accelerator power of 5 MW. Figure 9 shows the activation for the different materials after 1 day of beam off. The numbers shown in the figures represent the photon dose at 5 cm away from the highest hot spot of the collimator. Figure 10 shows the activation level after 7 days of beam off. As can be seen from the plots the activity of the collimator is not uniform across its length. The tungsten and the copper block will get quite active after one day but they will decay to a quite lower value after one week, while the mild steel will be still be very active. The reason for the different behaviour of the materials is due to the different half lives that drive the activation of the collimator. The tungsten activation after one day of beam off is mainly due to decays of ^{187}W (this isotopes has a half life of 23.72 hours), that will be mostly gone after one week, at that point the main decays will be due to ^{175}Hf and ^{182}Ta. The copper block activation after one day is driven by the decays of ^{64}Cu (half life of 12.7 hours) while after one week is mainly due to decays of ^{60}Co. For the mild steel the activation level is quite similar from 1 day to seven days since it is driven by the ^{60}Co decay that has a half life of more than five years.

5. Conclusions

Based on these simulations, it is recommended to use at most 30 cm of copper or tungsten at 16m around the vacuum pipe for NMX. It is important to underline that for the copper and tungsten in the simulations pure materials have been assumed. If instead of the pure material an alloy will be used, the presence of impurities could change significantly the activation results.

References

[1] https://europeanspallationsource.se/instruments/nmx

[2] T. Goorley, et. al., "Features of MCNP6", Supercomputing in Nuclear Applications and Monte Carlo 2013, Paris, Oct 27-31, LA-UR-13-28114 (2013).

[3] S. Ansell, "CombLayer: A fast parametric MCNP(X) model constructor", Proceedings of the 21st Meeting of the International Collaboration on Advanced Neutron Sources, Mito, Japan, Feb. 2016.

[4] W. B. Wilson, S. T. Cowell, T. R. England, A. C. Hayes & P. Moller: A Manual for CINDER'90 Version 07.4 Codes and Data, LA-UR-07-8412

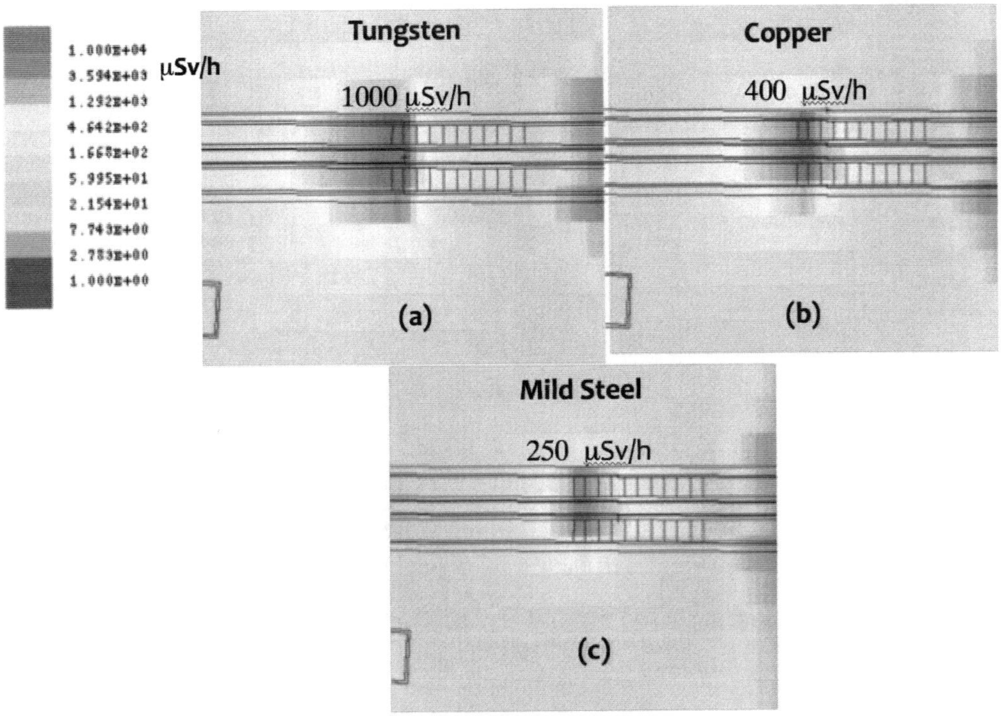

Figure 9: Photon radiation dose rates for the NMX collimator block located around 16 m from the moderator after one day of beam off. (a) Tungsten (b) Copper and (c) Mild Steel (see table 2).

IOP Conf. Series: Journal of Physics: Conf. Series **1046** (2018) 012010 doi:10.1088/1742-6596/1046/1/012010

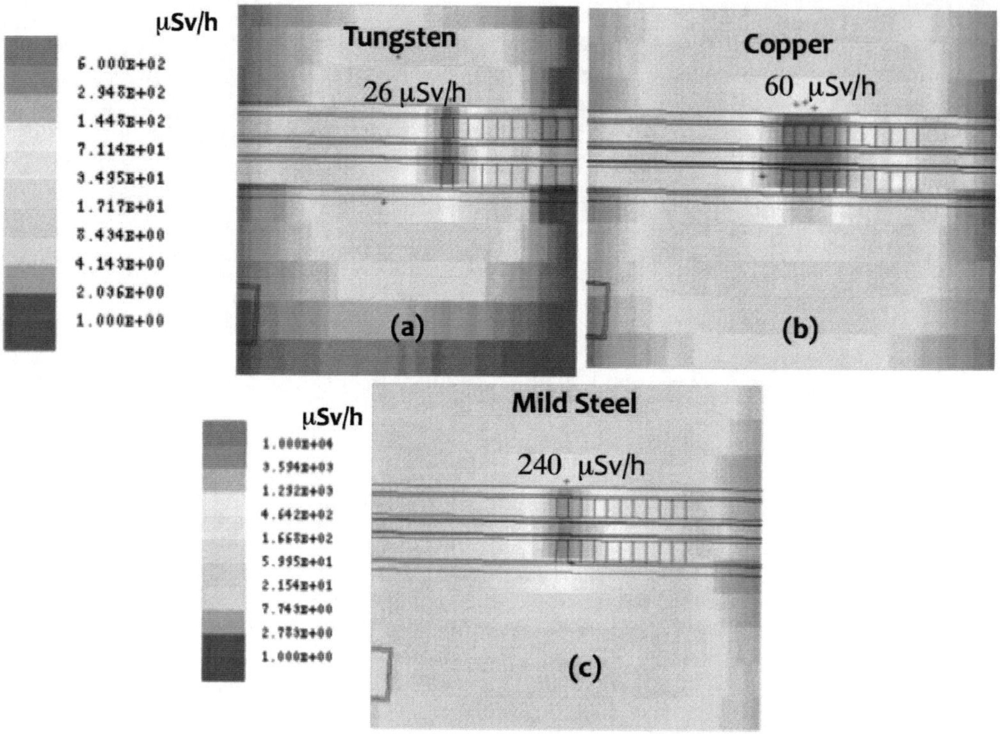

Figure 10: Photon radiation dose rates for the NMX collimator block located around 16 m from the moderator after seven days of beam off. (a) Tungsten (b) Copper (c) and Mild Steel (see table 2).

IOP Conf. Series: Journal of Physics: Conf. Series **1046** (2018) 012010 doi:10.1088/1742-6596/1046/1/012010

Table 2: Mild Steel composition used in the simulations .

Element	Atomic Mass	atom (%)
Mild Steel		
B	10	3.99×10^{-5}
C	12	1.20×10^{-3}
N	14	8.00×10^{-5}
Al	27	1.27×10^{-3}
Si	28	9.96×10^{-4}
P	31	5.00×10^{-4}
S	32	4.00×10^{-4}
Ti	47	1.00×10^{-4}
V	51	1.00×10^{-4}
Cr	52	1.14×10^{-3}
Fe	54	0.986
Mn	55	4.00×10^{-3}
Ni	58	1.00×10^{-3}
Co	59	2.00×10^{-4}
Cu	63	1.50×10^{-3}
Mo	92	1.00×10^{-3}
Nb	93	1.00×10^{-4}
Sn	118	9.93×10^{-5}

IOP Conf. Series: Journal of Physics: Conf. Series **1046** (2018) 012011 doi:10.1088/1742-6596/1046/1/012011

Shielding activated return water at the ESS

Esben Klinkby[1,2], Günter Muhrer[1], H. Carlsson[1] and Björn Eriksson[1]

1) European Spallation Source ESS, Box 176, S-221 00 Lund, Sweden

2) DTU Nutech, Technical University of Denmark, Frederiksborgvej 399, DK-4000 Roskilde, Denmark

E-mail: esbe@dtu.dk

Abstract. ESS utilises water both for moderating neutrons to thermal energies, as well as to cool beryllium - and steel reflectors, the shielding and plugs. This means that the water, in separate loops, will be subject to a significant proton and neutron irradiation causing the water to activate. After irradiation, the water is led to delay tanks situated inside the target building. Before returned to the target monolith $\sim$ 10% is led to the ion exchanger.

This paper aims at determining the shielding required to ensure that the biological dose-rate requirements in the target building and neighbouring instrument halls are met during operation of facility.

1. Introduction

The construction of the European Spallation Source (ESS) is ongoing in Lund, Sweden. At ESS it is necessary to operate water loops to remove heat from the thermal moderators and as well as the beryllium- and the steel reflectors, shielding blocks and plugs. The water in each of these circuits is exposed to a very high level of radiation, from as well protons and neutrons at various energies. This causes activation of the water and impurities therein. In this work the measures for safe handling of the return water are investigated. The work presented here focusses on operations (i.e. proton driver is on) as opposed to maintenance, which require a separate dedicated study (ongoing).

2. Methods and Modelling

The radiation shielding calculations are performed following the ESS Procedure for designing shielding for safety [1]. Monte Carlo radiation transport calculations are performed with the MCNPX-2.7 code [2]. To determine the source term the MCNPX master model of the ESS target-moderator-reflector system (v. 2.000) is run with NPS=1E6 to produce a *F4* tally and *histp* file of a cell which includes the combined water of the moderators and reflectors. Impurities are added to the water according to [3]. Next, based on the produced *histp* file and *F4* tally, a CINDER'90 (v. 1.05)[4, 6] activation calculation is carried out using as irradiation scenario the expected ESS operations schedule for two years of full power operations.

As seen in figure 1 the piping leading the water through the various systems is complex, and is not modelled in detail in this study. Instead, focus is put on the moderator loop, which is the most challenging in terms of activation. Approximate water volume of various fractions of the pipe circuits is used to define volumetric source terms (scaling the full source term), at various cooling times to account for the water flow. The cooling times considered fall in two categories:

Isotope	Half-life [s] [s]	Decay mode	Time step [0s]	Time step [90s]
^{3}H	3.89E8	β	621.3	621.3
^{7}Be	4.61E6	EC$\beta+\gamma[477keV]$	445.1	445.1
^{11}C	1223	β	1473	1400
^{14}O	70.6	$\beta+\gamma[2.3MeV]$	92.3	38.5
^{15}O	122	β	4511	2723
^{16}N	7.1	$\beta+\gamma[6.1MeV]$	4462	0.8
Total			12380	7161

Table 1: Total activity in Curies. Only isotopes contributing in excess of 10Cu at 60 s cooling are listed, but all are included in the total sum.

10*s* relevant to describe water flowing in pipes toward the delay tanks and 10 s - 90 s relevant for water in the delay tank.
To determine the required shielding, the radiation zoning of ESS is applied [5]. Both in the case of the moderator exhaust pipe as well the for the delay tanks, the limiting factor driving the shielding thickness, is the 3 μSi/h requirement of the instrument hall - see figure 1.

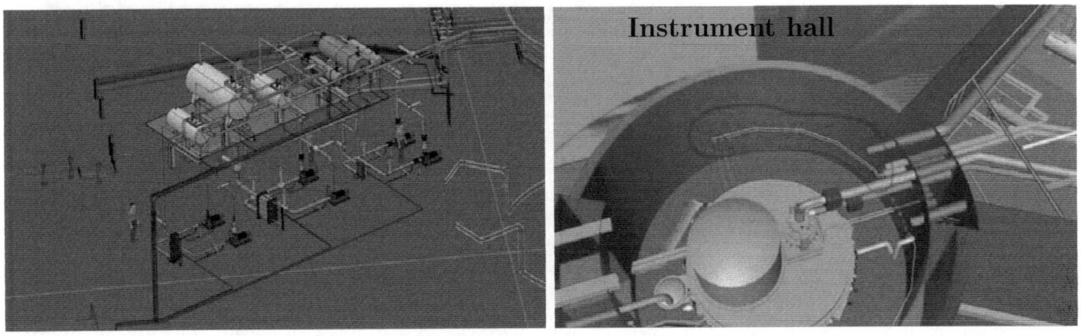

Figure 1: Left: CAD drawing of triangular rooms, showing the three orange delay tanks and connecting pipes. Right: connections cell, showing the exhaust pipe mounted on the wall toward the instrument hall.

For each cooling time considered, a source term is prepared (SDEF card) by executing the *gamma script* [6]. Next, a gamma transport simulation is carried out using MCNPX based on a simplified geometry of the target building and surroundings (next section). To determine dose-rate maps, in each configuration, $\geq$1E8 gammas are simulated, and the biological dose-rates are calculated using ICRP-116 gamma flux-to-dose conversion factors [7]. Importance biasing is implemented to ensure efficient calculations of radiation transport through the shielding.

3. Results

Table 1 lists the isotopes giving the largest contributions to the total activity at selected cooling times. As expected ^{16}N (^{16}O$(n,p)\rightarrow^{16}$N) is the main gamma emitter at short cooling times whereas ^{9}Be plays this role at longer cooling times. The combination of the very energetic gammas of ^{16}N with the fact that it's 7.1 s lifetime is of the same order as the time it takes the water to reach the connections cell from the moderator, makes this isotope particular challenging.

IOP Conf. Series: Journal of Physics: Conf. Series **1046** (2018) 012011 doi:10.1088/1742-6596/1046/1/012011

3.1. Shielding pipes in the connections cell

The pipes submerge from the protection steel of the target monolith in the connections cell and are routed toward the delay tanks in the triangular room as shown in figure 1(right). To describe the water source in the pipe following the circumference of the room, a cooling time of 10 s is assumed, accounting for the time it takes for the water to flow there from the moderators, situated 4 m below the floor.

The water volume present in the pipes correspond to about 1/4 of that present in the moderators, wherefore the source term is re-scaled by this factor. The resulting biological dose-rate maps are shown in Figure 2, for different shielding configurations as explained in the figure caption.

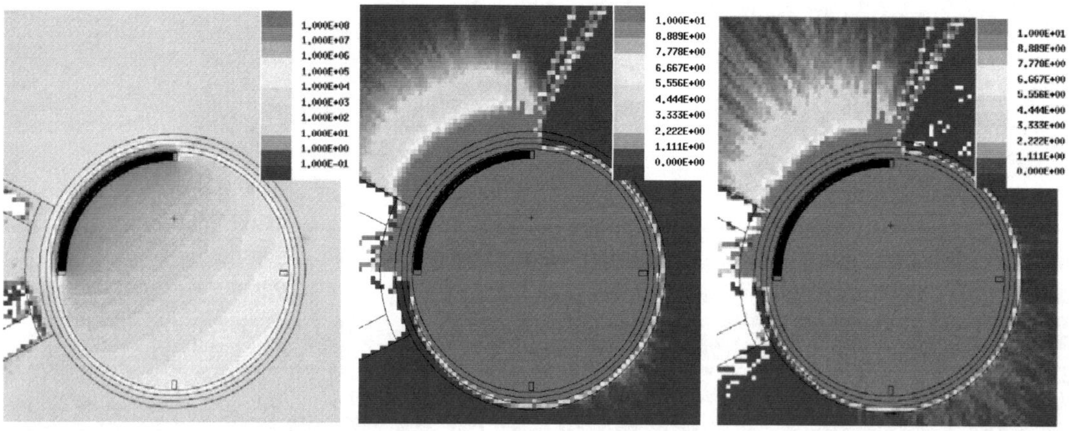

Figure 2: Biological dose-rate in μSv/h due to the delay tank inlet pipe. Shielding thickness: 0 cm (left), 12 cm lead (middle) and 6 cm lead + heavy concrete(right). Note the difference in scales: left-hand figure: $10^{-1} - 10^{8}\mu$Sv/h, middle and right-hand figures: 0-10μSv/h.

As seen in the left-hand insert of this figure, the challenge of shielding the exhaust water is significant. Two options are proposed, which provide sufficient shielding: either the pipe is surrounded by 12 cm of lead, or the composition of the wall towards the instrument hall is changed from regular concrete (ρ =2.35 g/cm^3) to heavy concrete (ρ =3.85 g/cm^3) in which case equivalent shielding is reached using 6 cm lead. At present it is being evaluated which of the two options is more practical and economic.

3.2. Delay tank shielding

The water circuits, of course, are not exactly steady. Moreover the timescales relevant to the water flow are similar to the half-life of ^{16}N, which according to table 1 is main contributer to the activity on a one-minute timescale. To accurately model the source term to be used for delay tank shielding calculations, the following observations are made:

- Moderator water content: 9L
- Moderator exhaust speed: 0.6 L/s (or 2 m/s) so the average cooling time, at the time of exiting the moderator: 7.5 s.
- Pipe length: ∼10 m. ∼5 m vertical and ∼5 m horizontal

Thus, *SDEF* cards are prepared corresponding to cooling times between 7.5 s + $\frac{10\text{ m}}{2\text{m/s}} \approx 13$ s and 37 s in 9 equidistant steps - see table 2.

Cooling time [s]	13	16	19	22	25	28	31	34	37
Source weight [%]	12	12	12	12	12	12	12	12	4

Table 2: Source definitions used to model the delay tank.

The resulting dose-rate map due to the presence of activated primary coolant water in the delay tank is shown in figure 3. The dose-rate requirements in the instrument hall are met only after introducing 16 cm of lead shielding. It should be noted that the delay tank need shielding on all horizontal directions, though the shielding requirements on the side facing the center of the room, can be reduced by approximately two orders of magnitude due to the geometrical dilution between the delay tank and the instrument hall on the opposite side of the connection cell.

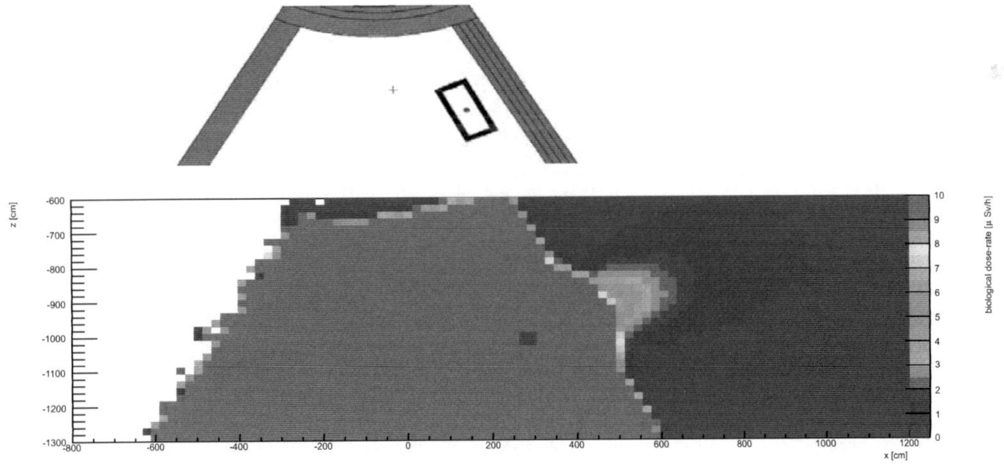

Figure 3: Biological dose-rate in μSv/h due to the delay tank - the geometry used in the biological dose-rate map (below), is indicated above. Shielding thickness: 16 cm lead.

4. Conclusions

The shielding requirements of the primary coolant return water is studied using a MCNPX neutron and gamma transport calculations in combination with CINDER'90. Immediately

IOP Conf. Series: Journal of Physics: Conf. Series **1046** (2018) 012011 doi:10.1088/1742-6596/1046/1/012011

after returning from the moderators/reflectors ^{16}N is the most challenging isotope, while after the cooling for 90 s in the delay tanks, ^{9}Be claims this role. For irradiated water present in the connections cell, the most challenging biological dose-rate limit to be met is that of the instrument hall. To meet the 3μSv/h limit, the following shielding must be installed:

- 12cm lead around delay tank inlet pipe
- 16cm lead between delay tanks and instrument hall

References

[1] ESS Procedure for designing shielding for safety, ESS-0019931
[2] Waters L. S. and others. The MCNPX Monte Carlo radiation transport code. *AIP Conf.Proc.* **896**, 81-90 (2007)
[3] ESS-0045043. Sheet: 2.Input and output data
[4] W. B. Wilson, S. T. Cowell, T. R. England, A. C. Hayes & P. Moller, A Manual for CINDER'90 Version 07.4 Codes and Data, LA-UR-07-8412 (December 2007, Version 07.4.2 updated March 2008).
[5] Radiation zoning for D02, ESS-0054402
[6] Franz X. Gallmeier and Michael Wohlmuther, ACTIVATION SCRIPT VERSION 1.0 USER GUIDE, ORNL/TM-2008/031 (August 18, 2008).
[7] Conversion Coefficients for Radiological Protection Quantities for External Radiation Exposures. ICRP Publication 116. Ann. ICRP 40(2-5), 2010

Radionuclide production and dose rate estimation during the commissioning of the W-Ta spallation target

Q Z Yu[1] and T J Liang[1]
[1] Beijing National Laboratory for Condensed Matter Physics, Institute of Physics, Chinese Academy of Sciences, Beijing 100190, China

qzhyu@iphy.ac.cn

Abstract. China Spallation Neutron Source (CSNS) is intended to begin operation in 2018. CSNS is an accelerator-base multidisciplinary user facility. The pulsed neutrons are produced by a 1.6GeV short-pulsed proton beam impinging on a W-Ta spallation target, at a beam power of100 kW and a repetition rate of 25 Hz. 20 neutron beam lines are extracted for the neutron scattering and neutron irradiation research. During the commissioning and maintenance scenarios, the gamma rays induced from the W-Ta target can cause the dose threat to the personal and the environment. In this paper, the gamma dose rate distributions for the W-Ta spallation are calculated, based on the engineering model of the target-moderator-reflector system. The shipping cask is analyzed to satisfy the dose rate limit that less than 2 mSv/h at the surface of the shipping cask. All calculations are performed by the Monte carlo code MCNPX2.5 and the activation code CINDER'90.

1. Introduction

The China Spallation Neutron Source (CSNS), currently under construction in Dongguan, Guangdong Province, will use a 1.6 GeV proton beam of 100 kW power to produce neutrons in a tantalum (Ta)-cladded tungsten (W) target [1,2]. There are approximately 30 neutrons released per incident proton, resulting in a total neutron source of $\sim 1\times10^{16}$ n/s. 20 neutron beam-lines are designed to export the produced neutrons for the experimental measurement [3,4]. But nonetheless, only a small fraction of these neutrons is successfully directed out of the target station. The major produced neutrons slow down and eventually get absorbed in the spallation target, the moderators, the reflectors and even in the biologic shielding, causing intense radiation damage and activation of these components [5-7]. Radiation damage caused by these produced neutrons, as well as that caused by the primary protons and the secondary produced particles, limits the life time of these components [8,9]. Some of these components such as the W-Ta spallation target and the moderators, need to be maintained or replaced regularly. Since these components are highly activated and can emit intensive decay gamma-rays, shipping casks used for transporting off-site of the damaged components are needed to ensure the personnel safety. Once the radiation assessment for the maintenance scenarios of CSNS inner reflector plug was performed and analyzed [10]. The dose rate distributions and the shipping casks used to transport these activated components are also calculated and designed.

In this paper, we are aim to present the gamma dose rate distribution of the W-Ta spllation target irradiated by a high-energy proton accelerator. The thickness of the shipping cask used to transport the spallation target is calculated. The induced residual nuclei and the decay gamma rays are also analyzed. All these calculations are based on the target-moderator-reflector (TMR) engineering model.

2. Calculation code and method

To calculate the gamma dose rate and the shipping cask for the W-Ta spallation target, the Monte carlo code MCNPX2.5 [11], the activation code CINDER'90 [12], the Activation Script and the Gamma-source Script [13] were used. The calculation sequence is illustrated as follows. Firstly, MCNPX2.5 is used to simulate the spallation reaction taken place in the TMR system. Bertini intra-nuclear cascade model coupled with Rutherford-Appleton Laboratory (RAL) evaporation-fission model was used to determine the spallation products for high energy interactions (E>20MeV). The evaluated nuclear data library ENDF-B/VI was used to calculate the low energy neutron (E<20MeV) fluxes in a 63 energy group structure. Secondly, the activation code CINDER'90, which solves the Bateman equations and deduces the time dependence of isotope buildup and decay, was used to calculate the activity, decay heat and induced gamma ray in a 25 energy group structure. Thirdly, an Activation Script provides the interface between MCNPX2.5 and CINDER'90, providing the residual gamma rays in the interested cells, corresponding to an irradiation time and a cooling time. And a Gamma Source Script was used to extract the induced gamma ray information as a new source file, including each cell number, intensity, spectrum, and space position. Lastly, MCNP4C was used to transport these induced gamma rays, for the gamma dose rate and shipping cask calculations that are aimed to perform in this paper.

3. Geometry of the 100 kW W-Ta target

The geometry models of the 100 kW CSNS W-Ta target and the TMR system are shown in Figure 1 (a) and (b), which are near to the engineering models [2]. This W-Ta spallation target is 17 cm wide, 7 cm high and 66.86 cm long. It consists of 11 W plates, each is surrounded by 0.3 mm thickness Ta claddings. The total length of the W plates is 67.1 cm, ensuring complete stopping of the 1.6 GeV protons in the target. There are 12 cooling water channels with 1.2 mm width between these W-Ta plates, which are used to remove the deposited nuclear heat in the spallation target. On the four sides of the W-Ta target, there are four water channels guiding the cooling water flows. SS316 steel with 1.0 cm thickness, is used as the target vessel. On the nose of the target, a 0.25 mm thickness $Cr:Al_2O_3$ luminescent coating is used to measure the incident proton beam profiles in real time.

The coupled hydrogen moderator (CHM) is arranged under the spallation target, while the decoupled water moderator (DWM) and the decoupled and poisoned hydrogen moderator (DPHM) are on the spallation target. Around the spallation target and the three moderators, a 27 cm in radius, 70 cm high beryllium (Be) cylinder is used as the inner reflector. Outside the Be reflector, a 47 cm in radius, 100 cm high iron (Fe) cylinder is used as the outer reflector. 3 cm thinkness aluminum is used as the container of the reflectors. Outside the Be/Fe reflectors is the biologic shielding. Although Fig. 1 (a) and (b) are the MCNPX models for the W-Ta spallation target and the TMR system, detailed geometries of each component can be seen, including the claddings, coatings, pipes, gaps, void spaces and so on.

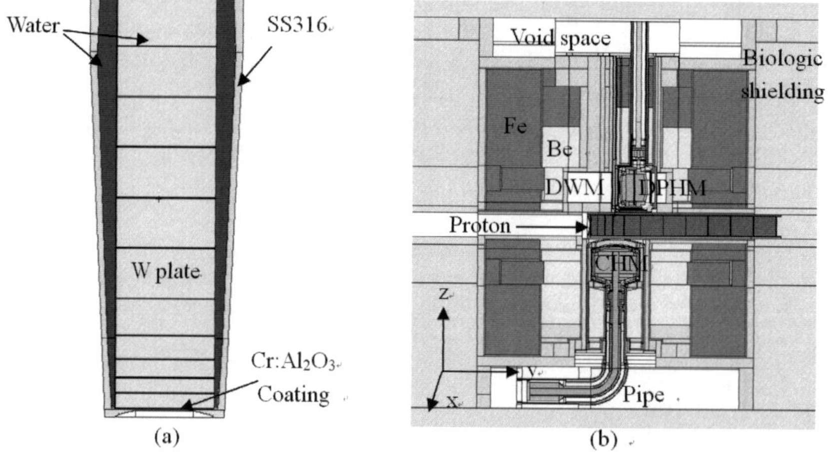

Fig. 1 MCNPX models for (a) W-Ta spallation target and (b) TMR system at CSNS.

All calculations are based on a 1.6 GeV, 100 kW (3.9×10^{14} p/s), 25 Hz proton beam irradiation the W-Ta spallation target. The footprint of the proton beam profile is 16 cm width and 6 cm high, with double Gaussian distributions of 9.42 cm and 3.53 cm full width at half maximum (FWHM) in x and z directions. The target is irradiated with full proton beam power for 5 years (5000 h/y). After 7 days cooling by air, it will be taken to the hot cell for replacement, or transported by a shipping cask for maintenance. The acceptable dose rate at the surface of the shipping cask is limited to 2 mSv/h.

4. Calculation results

When high-energy proton beam irradiates a W-Ta target, intense spallation reaction takes place. Various high-energy particles (protons, neutrons, pions, etc.), low-energy particles (protons, neutrons, deuterons, tritons, 3He, alpha, etc.), and a wide range of residual nuclei are produced. The radioactive inventories and the induced gamma rays, will impact the personnel safety and the environment. The problems include when the W-Ta spallation target has to be maintained, transported, decommissioned and finally disposed. Detailed calculations and analyses about the residual nuclides, decay gamma rays, gamma dose rate distributions and shipping cask are shown below.

4.1 The residual nuclei

During the intense spallation process, a wide range of residual nuclei are produced in the W-Ta spallaiton target. The residual nuclei production rates in W plates, Ta claddings and SS316 containers are presented in Figure 2 (a), (b) and (c), respectively. The same color bars are used for the production rate comparisons. It can be seen that for each components, the produced residual nuclei cover from ^{1}H to the parent nucleus element. For an example, in the W plates, there are about 1500 nuclides produced during the spallation process. Some of the light nuclides with charge number less than 10, are

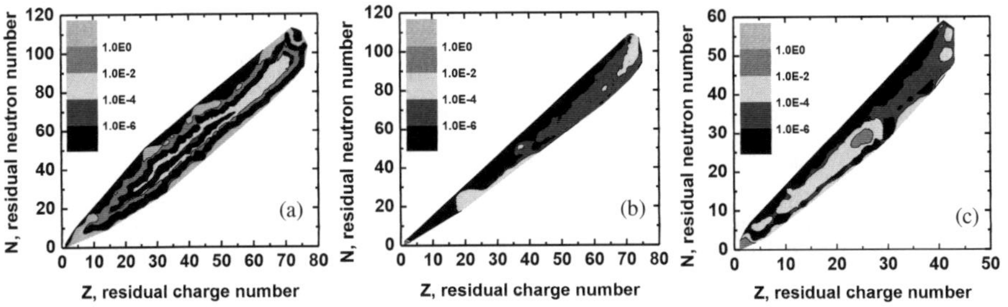

Fig. 2 (Color online) The produced residual nuclei in (a) W plates, (b) Ta claddings and (c) SS316 containers, as functions of residual charge number Z and residual neutron numbers N.

produced via the evaporation process. A large number of medium nuclides with charge number between 30 and 50 are produced via the fission process. The high charge number nuclides, which are near the parents nuclides (Z=74), are produced during the cascades process. For the residual nuclides produced in Ta claddings and SS316 containers, their production mechanisms are similar, while the production rates are much lower than that in W plates.

Most of these residual nuclides are radioactive, releasing different energies of electrons, α, β particles and decay gamma rays, reaching their stable states eventually. Since the decay gamma rays are not charged but with high energies, it should be taken more attentions.

4.2 The induced gamma rays

IOP Conf. Series: Journal of Physics: Conf. Series **1046** (2018) 012012 doi:10.1088/1742-6596/1046/1/012012

As mentioned above, the produced residual nuclides in W-Ta spallation target are highly radioactive. These radio-nuclides can induce the decay gamma rays. We show here the decay gamma rays in the W plates as an example. Figure 3 (a) shows the gamma intensity in each W plate, corresponding to 5 years irradiation and 7 days cooling. It can be seen that the most intense gamma ray presented in the first W plate, reaching to $\sim 8\times10^{11}$ p/s/cm^3. The gamma intensity exceed $\sim 10^{11}$ p/s/cm^3 for the 1~6 plates. It decreases rapidly along the target to $\sim 8\times10^{8}$ p/s/cm^3 at the last W plate. Figure 3(b) shows the induced gamma ray spectra in the 1, 4, 7, 10 W plates, respectively. The energy bin is according to the 25 energy-group that set in CINDER'90. It shows that most of the gamma rays are with energies less than 1 MeV, while a considerable portion has energy at several MeV. These high-energy gamma rays, can transport in the W-Ta spallation target to the outside region.

4.3 Gamma dose rate distribution and shipping cask

Although the self-shielding effect in the W-Ta target, the high-energy gamma rays can penetrate

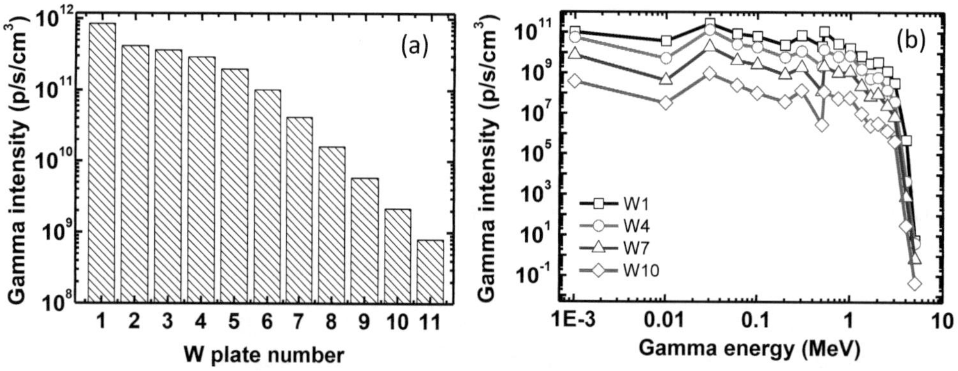

Fig. 3 (a) Induced gamma ray intensity in each W plate and (b) spectra in 1, 4,7,10 W plates, corresponding to 5 years irradiation and 7 days cooling.

the target itself to the out region, causing the gamma dose rate threat to the personal and the environment. A shipping cask is needed to shield the target during its maintenance or transportation. Figure 4 (a) is a MCNP4C geometry model of a Pb shipping cask for the W-Ta spallation target. A 2 cm air layer near the target is designed for the engineering consideration. It means that the Pb layIn this calculation model, each Pb layer is set as 2 cm thickness, for the use of the variance reduction of the gamma rays. Figure 4 (b) is the gamma dose rate map for the W-Ta spallation target, in the x-y plane at z=0. It can be seen that in the W-Ta target, the most intense dose rate can reach $\sim 10^6$ mSv/h. At the surface of the target, the dose rate is $\sim 10^5$ mSv/h. This means that the induced gamma rays emitted from the activated W-Ta spallation target, can cause serious radiation threat to the surroundings. The shielding effect of Pb cask is needed during its close range operation. We can see the dose rate varies distinctly in the Pb layers. At the surface of the Pb cask, the dose rate decreases to 1.0 mSv/h.

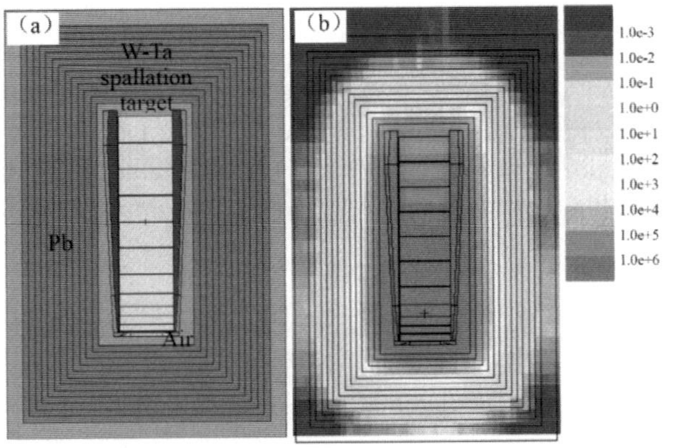

Fig. 4 (Color online) (a) MCNPX model of a Pb shipping cask for the W-Ta spallation target and (b) the corresponding gamma dose rate distributions.

For better analyzing the gamma dose rate distribution and the needed Pb thickness, we calculate the dose rate along x, y and z directions. Figure 5(a) shows the dose rate along x axis, at y from -8.95 cm to -6.65 cm (the thickness of the first W plate) and z=0. Figure 5 (b) shows the dose rate along y axis, at x from -5 cm to 5 cm and z=0. Figure 5 (c) shows the dose rate along z axis, at x from -5 cm to 5 cm and y from -8.95 cm to -6.65 cm. It can be seen that the dose rate in the W-Ta spallation target is higher than 1.0×10^6 mSv/h. The dose rate decreases exponentially in the Pb cask. Results show that at both x directions, 19.0 cm thickness of Pb is needed to decrease dose rate to 2 mSv/h at the surface of the Pb cask. At y direction, the dose rate variations in the W-Ta target can be seen. In front part of the spallation target, the dose rate exceeds 1.0×10^6 mSv/h. While at the back of the target, it is about 1.0×10^5 mSv/h. This causes the asymmetric dose rate distributions and different Pb shielding thickness. Calculation shows that 19 cm thickness Pb layer is needed in front of the target (-y direction), while 15.4 cm Pb layer is required at the backend of the target (+y direction). At both z directions, the dose rate distributions are symmetric. 21 cm thickness Pb is needed to satisfy the dose rate limitation. By calculating the dose rate distributions, the dimensions of a shipping cask can be determined.

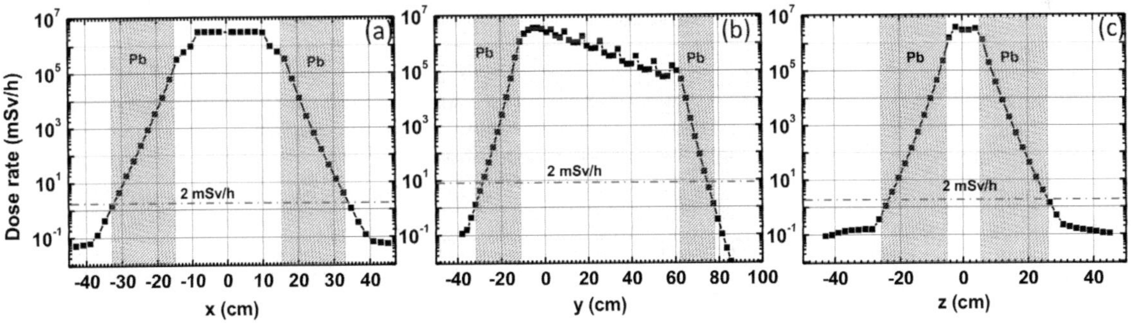

Fig. 5 Dose rate distribution along x axis (a), y axis (b) and z axis (c). The dose rate decreases exponentially in the Pb cask (gray line region). 19.0 cm thickness of Pb is needed at both x directions. 19 cm Pb layer is needed at –y direction, 15.4 cm Pb layer is needed at +y direction. 21.0 cm thickness of Pb is needed at both z directions.

5. Conclusion

The CSNS facility is intended to be operation in 2018. Due to the W-Ta spallation target is the most activated components in the target station, its commissioning is essential for the progress of the whole project.

Calculations have been carried out to determine the gamma dose rate around the W-Ta spallation target, which was irradiated by a 1.6 GeV proton beam with 100 kW beam power. Due to the intense spallation reactions, various particles and radio-nuclides are produced. The decay gamma rays induced from the radioactive inventories, can transport out of the W-Ta target itself, resulting the dose rate distributions around the target. It shows that the dose rate can reach ~10^6 mSv/h in the center of the W-Ta spallation target. At the surface of the target, the dose rate is at the level of ~10^5 mSv/h. Such kind of dose rate can cause a serious threat to the personnel during the close operation of the activated target. It also means that after a long time irradiation of the W-Ta spallation target, a shipping cask is needed to decrease the dose level to the safety range. Calculation shows that 19 cm thickness Pb shipping cask is needed to decrease the dose rate to 2 mSv/h limitation at the surface. These results and analysis are essential for the commissioning of the W-Ta spallation target.

Acknowledge

This work was supported by the National Science Foundation of China (Grant Nos. 11575289 and 11174358).

References

[1] http://csns.ihep.ac.cn/english/index.htm.
[2] Wei J, Chen H S, Chen Y W., et al., China Spallation Neutron Source: design, R&D, and outlook, Nucl. Instr. Meth. Phys. Res., A 600 (2009) 10.
[3] Wang F W, Liang T J., Yin W, Yu Q Z, et al., Physical design of target station and neutron instruments for China Spallation Neutron Source, Sci. Chin. Phys. Mech. Astron. 56 (2013) 2410.
[4] Yu Q Z, Yin W, Liang T J, A physical design of a neutron irradiation spectrometer at CSNS facility, 11th International Topical Meeting on Nuclear Applications of Accelerators, AccApp (2013)55-59
[5] Yu Q Z, Liang T J and Yin W, Activity and radiation protection studies for the W-Ta target of CSNS, Radiat. Prot. Dosimetry, 136 (2009) 216–221.
[6] Yin W, Liang T R, Shen F, Wang S L, Yu Q Z and Liang T J, Shielding calculation for CSNS target station, Progress in Nuclear Science and Technology, 4(2014) 210.
[7] Yu Q Z, Liang T J, Yin W and Yan Q W, Induced radioactivity in components of CSNS target station, PSI Proceedings, (2009) 164.
[8] Yu Q Z, Yin W, Liang T J, Calculation and analysis of DPA in the main components of CSNS target station, Acta Phys. Sin. 60 (2011) 052501-8.
[9] Yin W, Yu Q Z, Lu Y L, Wang S L, Tong J F and Liang T J, The expected radiation damage of CSNS target, J. Nucl. Mater., 39 (2012) 431.
[10] Yu Q Z, Hu Z L, Zhou B, et al., The radiation assessment for the maintenance scenarios of CSNS inner reflector plug, Progress in Nuclear Science and Technology, 4 (2014) 376-379.
[11] MCNP/MCNPX CCC-730, Monte Carlo N-Particle Transport Code System Including MCNPX 2.5.0 and Data libraries, (2006).
[12] Wilson W B, England T R and Van R K A, Status of CINDER'90 Codes and Dat, LA-UR-99-316, Los Alamos National Laboratory, (1999).
[13] Gallmeier F X, Wilson W L, Wohlmuther M, et al., An environment using nuclear inventory codes in combination with the radiation transport code MCNPX for accelerator activation problems, AccApp07, Pocatello, Idaho, 2007, 207.

IOP Conf. Series: Journal of Physics: Conf. Series **1046** (2018) 012013 doi:10.1088/1742-6596/1046/1/012013

Calculations vs. measurements of remnant dose rates for SNS spent structures

I.I. Popova[1], F. X. Gallmeier[1], S. Trotter[1], M. Dayton[1]
ORNL[1], PO Box 2008, MS 6466, Oak Ridge, TN 37831-6474, US

popovai@ornl.gov

Abstract. *Residual dose rate measurements were conducted on target vessel #13 and proton beam window #5 after extraction from their service locations. These measurements were used to verify calculation methods of radionuclide inventory assessment that are typically performed for nuclear waste characterization and transportation of these structures. Neutronics analyses for predicting residual dose rates were carried out using the transport code MCNPX and the transmutation code CINDER90. For transport analyses complex and rigorous geometry model of the structures and their surrounding are applied. The neutronics analyses were carried out using Bertini and CEM high energy physics models for simulating particles interaction. Obtained preliminary calculational results were analysed and compared to the measured dose rates and overall are showing good agreement with in 40% in average.*

1. Introduction

The Spallation Neutron Source (SNS) in Oak Ridge, Tennessee, is an accelerator driven neutron scattering facility for materials research. Presently SNS is capable and approved to operate at 1.4 MW proton beam power incident on a mercury target with a proton beam energy of 1 GeV and 60 Hz repetition rate.

SNS target system components are periodically replaced because they reach their end-of-life due to radiation induced material damage. The target vessel, which houses the mercury target, is exchanged about two-three times per year and the proton beam window (PBW) is exchanged ones per two – three years. Both target vessel and PBW are exposed to severe radiation environment, and are replaced to avoid events of spills due to structural failure caused by excessive material degradation and faitigue.

All spent components must be safely removed, placed in a container for temporary on-site storage, and ultimately transported off-site to a nuclear waste disposal site.

In order to characterize and classify spent components, accurate estimates of radionuclide inventory are performed. A bounding case considering the maximum possible radionuclide inventory, based on a scenario with maximum possible irradiation exposed during the life-time, is established for the respective component. Using these data, the spent component is characterized and classified, and an appropriate container for temporary storage on-site and subsequent transport off-site is suggested.

[1] This manuscript has been authored by UT-Battelle, LLC under Contract No. DE-AC05-00OR22725 with the U.S. Department of Energy. The United States Government retains and the publisher, by accepting the article for publication, acknowledges that the United States Government retains a non-exclusive, paid-up, irrevocable, world-wide license to publish or reproduce the published form of this manuscript, or allow others to do so, for United States Government purposes. The Department of Energy will provide public access to these results of federally sponsored research in accordance with the DOE Public Access Plan (http://energy.gov/downloads/doe-public-access-plan).

Once the container has been selected, transport calculations for the Department of Transportation (DOT) package are performed to ensure that the transport package is compliant with transportation and waste management regulations.

Supporting documentation with radionuclide inventory and dose rate prediction for the time of the transportation is prepared for each shipment. Neutronics analyses are performed, assuming realistic irradiation history and decay case to ensure that the container/package, housing the structure, is compliant with the waste management regulations. Analyses are complicated due to geometry, multi-code usage and following data treatment.

To validate analyses, dose rates from the spent target vessel #13 and PBW module #5 were measured. Neutronics analyses were performed to calculate residual dose rates from both structures for the time of measurements. This work presents comparison of preliminary analyses vs. measurements of the residual dose rates.

2. Methods for neutronics analyses

Complete analyses for the spent component radionuclide inventory and dose rates are performed for each off-site shipment, containing target vessel or PBW module, with realistic irradiation history and for the time of the shipment. Because these analyses are occurring on regular basis, to facilitate the process, the automated script system, which runs the analyses and generates the report, for both target vessel and PBW was developed.

Full three-dimensional radiation transport calculations with the state-of-the-art Monte-Carlo code MCNPX Version 2.7.0 [1] and the latest as-built target station model including the PBW module (Figure 1) are performed to simulate the radiation environment. Specifically, isotope production rates due to spallation reactions and the below-20-MeV neutron fluxes in the 63 group CINDER90 group structure were calculated for target facility components of interest. The calculations were performed twice with the two different high energy physics models Bertini and CEM.

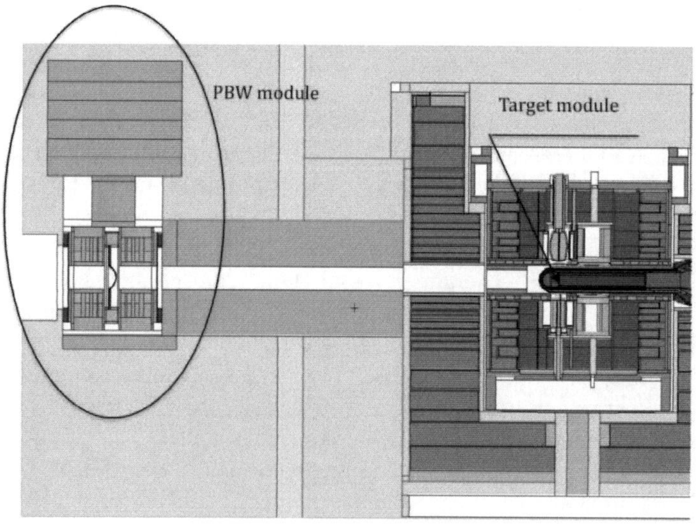

Figure 1. MCNPX model of target station, including PBW module, vertical cross section through the beam center line.

The obtained isotope reaction rates and neutron fluxes were extracted from the transport calculation output and were fed into the CINDER90 transmutation code [2] [5] using the standardized ACTIVATION_SCRIPT [3] to calculate the radionuclide inventory of the components. In order to obtain local distributions of radionuclide inventory and subsequent decay source terms, the components were subdivided into small pieces. The target vessel model was represented by 140 cells and the PBW model by 116 cells. The target vessel radionuclide inventory includes the target vessel, which is made from stainless steel 316L, and, as we assume, 200 g of activated mercury dispersed in the target together with 10% of the mercury radionuclide inventory (other than mercury, gold and noble gas isotopes) deposited on the mercury exposed steel piping and target structure of the mercury loop. The PBW module is comprised from an Inconel-718 window and is encased in a 316-stainless steel holder, and a water-cooled collimator assembly also of 316-stainless steel.

Realistic irradiation history and decay time, reconstructed from archived measured accumulated energy of SNS operations, are used in the analyses.

Decay gamma sources for a defined history of build-up and decay were extracted from each cell and compiled to a source term in MCNPX language by running the GAMMA_SOURCE_SCRIPT [4]. The decay gamma source was prepared for the time, when measurements of the dose rates occurred.

Then, the decay gamma source terms were utilized in photon transport calculations for both target vessel and PBW module. To simulate measurements, box-type volumes of 2x2x2cm^3 at the detector positions were defined around the components at the locations of the measurements. For the residual dose rate scoring in the detector volume, a track-length flux tally, was applied. Dose rates were obtained by folding fluxes with SNS specific flux to dose conversion coefficients [6].

3. Measurements

The specialized high-range radiation detection instrumentation Ludlum Model 9-7 with 9-7-BH Detector (Figure 4) was used for the measurements of residual dose rates. According to specification, the detector range is 0.01 – 19.99 kR/hr, with the resolution of 0.01 kR/h and the linearity of 10%. The active detector volume (ion chamber) is 7cm^3, with 2.54cm diameter and 1.64 cm thickness. Detector calibration was performed at ORNL with a Cs-137 source. Certificate of calibration was issued.

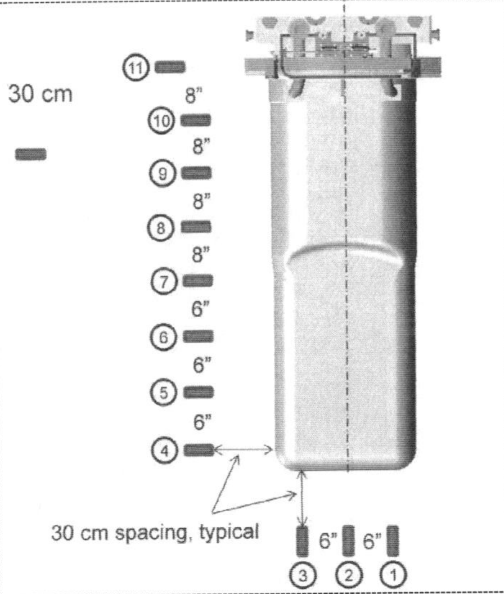

Figure 2. Detectors positioning around the target vessel #13.

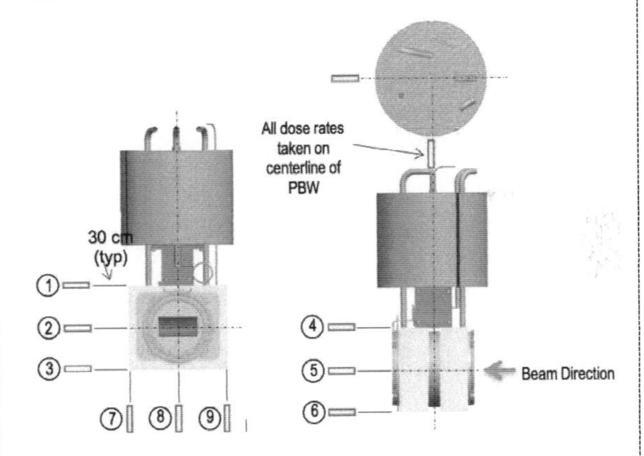

Figure 3. 2 Detectors positioning around the PBW module #5.

4. Results

The analyses for residual dose rates were performed for the time of measurements, which is 106 days for the target #13 and 80 days for the PBW module #5 after beam termination on target. The standard deviation for most of the calculated dose rate values was less than 1%, beside a few locations, where the standard deviation was slightly higher but lower that 4%.

4.1. Target

Service lifetime [8] for target vessel #13 is approximately 0.50 years in which it accumulated slightly more than 2588.55 MWh proton beam energy.

For the dose rate analyses the fluxes under 20 MeV and the high-energy radionuclide production rates were obtained using:

- MCNPX version 2.5 with Bertini model for high-energy particles interactions;
- MCNPX version 2.7 with Bertini model for high-energy particles interactions;
- MCNPX version 2.7 with CEM model for high-energy particles interactions.

Mercury, which is left over after the drainage from the target vessel, was modeled in two way – dispersed inside the target vessel, and as a paddle at of the target nose. Results are presented in Table 1. C/M refers to calculation over measurements ratio.

Overall the calculated dose rates are in a good agreement with the measured dose rates – within about 40%. Calculated dose rates at most locations have higher values compared to measured data. Using CEM model vs. Bertini model for high-energy particles physics does not show significant impact on calculations. Using reaction rates and fluxes generated by MCNPX version 2.7.0 compared by MCNPX version 2.5.0 improves the C/M comparison. Modeling 200g of mercury as a puddle vs. distributed mercury at lower density does not impact the dose rates significantly.

Table 1. Calculated dose rates vs measured dose rates for target vessel #13, mrem/h.

		Dispersed mercury						Mercury paddle			
Detector#	Measurements	MCNPX 2.5.0	C/M	MCNPX 2.7.0	C/M	MCNPX 2.7.0, CEM	C/M	MCNPX 2.5.0	C/M	MCNPX 2.7.0	C/M
1	2.04E+6	3.35E+6	1.64	2.94E+6	1.44	3.00E+6	1.47	2.83E+06	1.39	3.00E+06	1.47
2	2.42E+6	3.80E+6	1.57	3.28E+6	1.36	3.36E+6	1.39	3.19E+06	1.32	3.38E+06	1.40
3	1.96E+6	3.34E+6	1.70	2.95E+6	1.51	3.02E+6	1.54	2.86E+06	1.46	3.03E+06	1.54
4	1.40E+6	1.92E+6	1.37	1.71E+6	1.22	1.65E+6	1.18	1.66E+06	1.19	1.76E+06	1.26
5	1.60E+6	2.31E+6	1.44	2.05E+6	1.28	1.99E+6	1.24	2.00E+06	1.25	2.12E+06	1.32
6	1.45E+6	2.25E+6	1.55	2.00E+6	1.38	1.93E+6	1.33	1.95E+06	1.34	2.06E+06	1.42
7	1.09E+6	1.85E+6	1.70	1.63E+6	1.50	1.58E+6	1.45	1.59E+06	1.46	1.68E+06	1.54
8	6.60E+5	1.08E+6	1.64	9.89E+5	1.50	9.69E+5	1.47	9.95E+05	1.51	1.05E+06	1.59
9	3.70E+5	5.50E+5	1.49	5.12E+5	1.38	4.92E+5	1.33	5.15E+05	1.39	5.45E+05	1.47
10	2.20E+5	2.45E+5	1.12	2.69E+5	1.22	2.41E+5	1.10	2.67E+05	1.21	2.83E+05	1.29
11	1.60E+5	1.45E+5	0.91	1.57E+5	0.98	1.34E+5	0.84	1.60E+05	1.00	1.69E+05	1.05

4.2. PBW

For the residual dose rate analyses, the fluxes under 20 MeV and the reaction rates were obtained using MCNPX version 2.7 with the Bertini model for high-energy particles interactions. Results are presented at Table 2.

Table 2. Calculated dose rates vs measured dose rates for PBW #5, kRem/h.

Detector #	Measurements	Calculations	C/M
1	0.16	0.22	1.38
2	0.22	0.24	1.09
3	0.17	0.2	1.18
4	0.38	0.4	1.05
5	1.15	1.03	0.90
6	0.52	0.38	0.73
7	0.22	0.23	1.05
8	0.26	0.31	1.19
9	0.21	0.24	1.14

IOP Conf. Series: Journal of Physics: Conf. Series **1046** (2018) 012013 doi:10.1088/1742-6596/1046/1/012013

Most of the calculated results are within 20% of the measured data, and are generally in the higher side. There are two locations where the dose rates differ by 40%. This difference could be caused by uncertainty in the detector position during the measurements.

5. Conclusions

In order to validate neutronics analyses, measurements of dose rates from a spent target vessel and a PBW module were performed. Preliminary neutronics analyses were conducted to calculate residual dose rates from the target vessel and PBW for the time of the measurement. Overall calculated dose rates are in agreement with measured dose rate to within 40%, beside for a few locations that lie within 60%. We are pleased with the outcome of this validation effort considering the complexity of the geometry and involvement of different calculational codes.

6. Acknowledgment

ORNL is managed by UT-Battelle, LLC, under contract DE-AC05-00OR22725 for the Energy. This research was supported by the DOE Office of Science, Basic Energy Science, Scientific User Facilities.

References

[1] D. Pellowitz, ed., "MCNPX User's Manual, Version 2.7.0," LA-CP-11-00438, Los Alamos National Laboratory, Los Alamos, New Mexico (April 2011).

[2] D. Pellowitz, ed., "MCNPX User's Manual, Version 2.5.0," LA-CP-05-0369, Los Alamos National Laboratory, Los Alamos, New Mexico (April 2005).

[3] W. B. Wilson, S.T. Cowell, T. R. England, A.C.Hayes, P. Möller, A Manual for Cinder'90 Version07.4, LA-UR-07-8412, Los Alamos National Laboratory, Los Alamos, (2007).

[4] F. X. Gallmeier and M. Wohlmuther, Activation Script Version 1.0 User Guide, ORNL-TM-2008/031, Oak Ridge National Laboratory, August 2008

[5] M. Wohlmuther and F.X. Gallmeier, User Guide for the Gamma Source Perl Script 1.0, PSI-TM-85-08-02, Paul Scherrer Institute, July 2008.

[6] P. D. Ferguson, CINDER'90 for SNS Activation Studies, SNS-106100200-TR0142-R00, Oak Ridge National Laboratory, March 2006.

[7] I. Popova, Flux to Dose Conversion Factors, SNS-NFDD-NSD-TR-0001-R00, Oak Ridge National Laboratory (October 2009)

[8] 8. I.I. Popova, "Activation and Waste Handling of Target Vessel #13, SNS106100200-DA0074-R01, Oak Ridge National Laboratory (October 2009), August 2016.

IOP Conf. Series: Journal of Physics: Conf. Series **1046** (2018) 012014 doi:10.1088/1742-6596/1046/1/012014

Analysis of residual activity at the FRIB linear accelerator

M Kostin[1], G Bollen[1], D Georgobiani[1], P Grivins[1], R Lowrie[2] and R Ronningen[3]

[1] Facility for Rare Isotope Beams, Michigan State University, 640 South Shaw Lane, East Lansing, MI 48824, USA

[2] AECOM, 2131 South Centennial Avenue, Aiken, SC 29803-7680, USA

[3] National Superconducting Cyclotron Laboratory, Michigan State University, 640 South Shaw Lane, East Lansing, MI 48824, USA

kostin@frib.msu.edu

Abstract. The Facility for Rare Isotope Beams (FRIB) is an accelerator facility being established at Michigan State University (MSU). The facility will utilize a broad range of primary ion beams from ^{16}O to ^{238}U with a beam power of up to 400 kW and energy of 200 MeV/u for ^{238}U in its baseline configuration to produce rare isotopes. A possible facility upgrade will include an increase of the beam energy up to 400 MeV/u for ^{238}U and addition of new light ion beams down to ^{3}He and protons for Isotope Separation Online (ISOL) operations. The FRIB double-folded linear accelerator will accommodate several high-rate beam loss devices. The residual activity of these devices were analysed and results are presented in this work.

1. Introduction

There are several segments in the FRIB linear accelerator (figure 1): front-end that delivers the primary beam from ion sources down to the accelerator tunnel; Linear Segment 1 (LS1) where the primary beam is accelerated up to an energy of 20 MeV/u (16.5 MeV/u for uranium); Folding Segment 1 (FS1) where the primary beam is stripped of electrons and proper ion charge states are selected for further acceleration; Linear Segment 2 (LS2) where the primary beam is further accelerated (up to 149 MeV/u for uranium); Folding Segment 2 (FS2); Linear Segment 3 (LS3) which brings the beam energy up to 200 MeV/u for uranium; and Beam Delivery Section (BDS) which delivers the primary beam to the target hall. A number of devices with the beam loss rate above 100 W are found across the accelerator. Induced radioactivity of each of these devices will be discussed in dedicated sections. Analysis of a few other devices such as collimators and chicanes where beam loss rates are either lower or devices that can only be used with a limited set of specific beam types was not included in this work. Induced radioactivity of various utilities such as water and nitrogen and helium in the cryogenic systems, the activation of air and condensates are outside of the scope of this work.

The FRIB facility is expected to start the physics program in 2022. This is followed by a 5 year plan to ramp up to the full power. During this time period, measurements will be taken to understand actual beam loss rates and activity levels of the linear accelerator components. Future updates were planned in the design of the facility. In the simplest scenario, Linac Segment 3 can accommodate more superconducting accelerating structures to bringing the beam energy up to 400 MeV/u for uranium while keeping the same power of 400 kW. Alternate scenarios may include construction of a new Isotope Separation Online (ISOL) target station and a light ion injector, or replacing the accelerating

IOP Conf. Series: Journal of Physics: Conf. Series **1046** (2018) 012014 doi:10.1088/1742-6596/1046/1/012014

structures with new more efficient ones. In that case, the devices considered in this work are expected to be re-evaluated or redesigned.

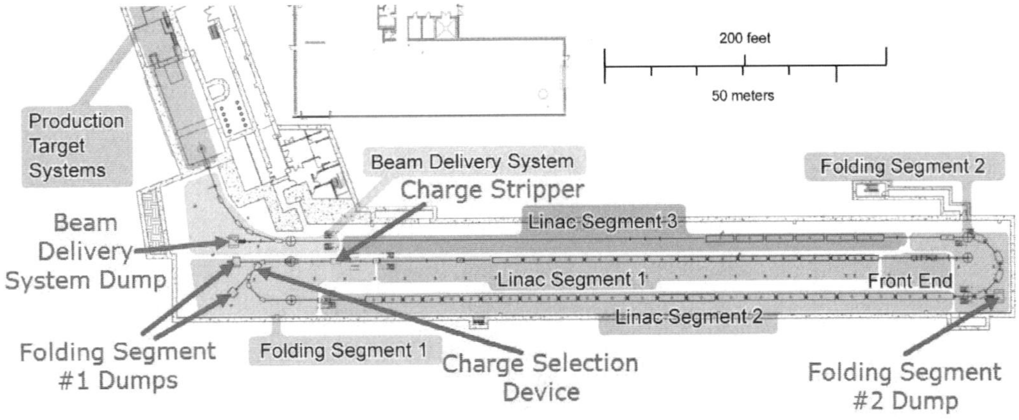

Figure 1. Schematics of the FRIB linear accelerator. The arrows indicate high-rate loss beam devices.

2. Rationale for device local shielding

Two main sources of irradiation inside the linac tunnel are normal operation uncontrolled beam losses and beam interactions with high-rate loss devices. A continuous beam loss rate of 1 W/m is assumed as baseline design input for the normal losses. In most cases, however, the activation of a beamline component is dominated by the local beam-device interactions if the loss rate in the device exceeds 100 W. The design goal is to provide enough local shielding for such a device so that the impact on the radiation environment from the local interactions would be comparable to that caused by the continuous 1 W/m beam losses. The three primary design considerations for the local shielding are the prompt dose to workers and public above the bulk tunnel shielding (6 m of concrete and soil), activation of the soil and the ground water around the tunnel, and the residual activation for the purpose of hands-on maintenance. Only the residual activity is of importance at Linac Segment 1 and Folding Segment 1 due to the low energy of the primary beam. The design goal for the local shielding from the standpoint of hands-on maintenance is to keep the dose rate from activated component below 5 mrem/h (50 μSv, 100 rem=1 Sv) whenever possible to prevent each area from being designated as a "Radiation Area" as specified by Nuclear Regulatory Commission (NRC) regulations [1] where additional controls are required to work in these areas. If the local shielding for a specific device is impractical due to the device size for example, then movable screens (temporary shielding) are considered to limit the exposure from the activated components to a manageable level.

3. Charge stripping device

The charge stripping device is located in Folding Segment 1. Its purpose is to strip the primary beam ions to higher charge states for more efficient acceleration in the downstream accelerating structures. All the beam types up to argon will be fully stripped. The uranium beam will have five charge states after the stripping device - 76, 77, 78, 79 and 80. A simple carbon foil-based device is considered for the initial operation while the beam intensities are still low. The baseline design will be based on liquid lithium curtain however.

A simplified model used in the calculation is shown in figure 2. There is a 1 mg/cm^2-thick lithium curtain, lithium storage vessel, argon gas cover volume, external argon storage vessel, and a steel containment vessel with two wall thicknesses considered – 0.5 inch and 1 inch. The radiologically bounding beam in this case is ^{18}O at 20 MeV/u. The full beam power at the stripping device is 40 kW. In addition, the contribution from the normal uncontrolled beam losses was evaluated. The activation of the device was calculated at a distance of 30 cm [1] from its walls assuming 30 y of irradiation and

IOP Conf. Series: Journal of Physics: Conf. Series **1046** (2018) 012014 doi:10.1088/1742-6596/1046/1/012014

4 h after shutdown. The programs PHITS **Error! Reference source not found.** and DCHAINSP-2001 [3] were used. The activities at other irradiation and decay times can be easily recalculated with existing tools based on known decay curves of common radionuclides. One of such tools is provided by MARS15 [4]-[7] in form of a matrix. Radionuclide production in argon and lithium were also evaluated as well as the consequences of the argon emissions, but these topics are outside of the scope of this work.

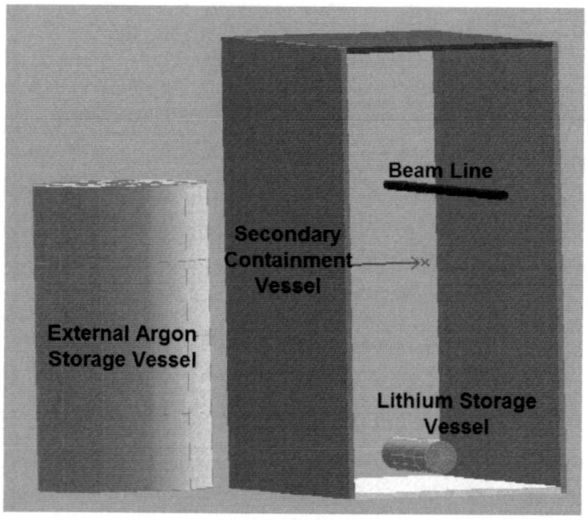

Figure 2. Geometry model used in radiation transport calculations for charge stripper activation studies. The beam line contains a liquid lithium target. The secondary containment vessel is filled with argon gas. The upstream and downstream enclosure walls are not shown.

It was found that the induced radioactivity at the stripping device is dominated by the beam-stripper interactions over the normal beam losses in all the locations except for an area upstream of the external argon reservoir. A typical dose rate distribution is shown in figure 3. Table 1 shows that, depending on the location and the wall thickness, the maximum dose rate is reaching 99.8 mrem/h. The FRIB physics program will require a mixture of various beams, therefore there is a reduction factor for the dose rate due to the use of other beams which will produce less intense radiation fields. Nonetheless, the dose rate in the area around the unshielded stripping device is expected to be above 5 mrem/h, therefore the area will be controlled as “Radiation Area”. Using additional local shielding is impractical in this case because of the substantial device size and because the dose rate is not much higher than 5 mrem/h. Portable shielding screens will be utilized to keep doses as low as reasonably achievable.

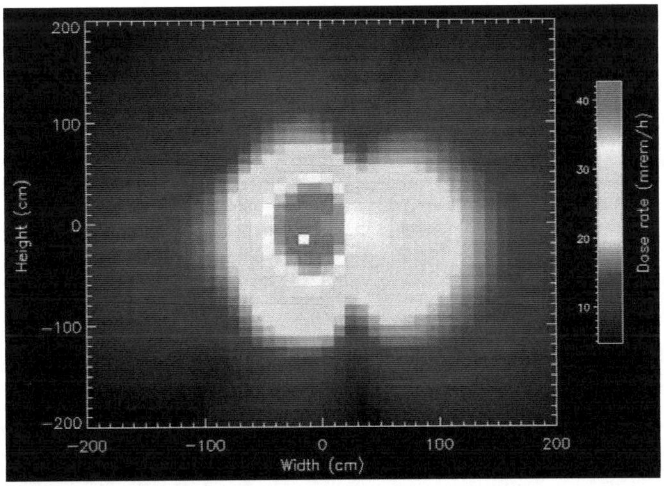

Figure 3. Residual dose rate distribution upstream of the secondary containment vessel.

Table 1. Maximum and average residual dose rates at various locations of the charge stripping device.

Location	Average dose rate (mrem/h), 0.5 inch vessel	Maximum dose rate (mrem/h), 0.5 inch vessel	Average dose rate (mrem/h), 1 inch vessel	Maximum dose rate (mrem/h), 1 inch vessel
Right side of device	18.6	49.3	12.5	33.1
Left side of containment vessel	27.2	99.8	19.8	74.4
Below containment vessel	13.6	34.5	9.3	21.8
Above containment vessel	16.7	47.5	11.4	32.9
Upstream of containment vessel	13.7	42.7	8.7	28.0
Downstream of containment vessel	20.5	75.3	17.0	72.5
Upstream of argon storage vessel	8.5	16.4	5.3	10.7
Left of argon storage vessel	9.5	18.1	6.3	12.7

4. Charge selection device

The main function of the charge selection device is to remove unwanted charge states from the beam after the charge stripper. At the moment of writing this contribution, a design for device capable of handling the full beam power has not been developed yet. A temporary device will be used during the beam commissioning and the initial operation until 10% of the full power is reached. Since the initial beam power will be limited, the temporary device is much simpler than a concept for the full power device. The temporary device will be equipped with movable water-cooled jaws as opposed to rotating graphite disks. Final full power design will be based on operational experience during power ramp-up to 400 kW. The low power charge selector will be based on the existing hardware with coper alloy jaws (see figure 4). The device will be enclosed in a 6 inch-thick steel box to mitigate the radiation streaming from the device (figure 5). Although the oxygen isotope beams are considered radiologically bounding, these beams will be fully stripped, and, therefore, the charge selection device will not be used with them. According to table 2, the lightest of the most commonly used beams will be ^{48}Ca, with 135 W deposited in the jaws. The heavier beams will be producing less intense radiation fields than ^{48}Ca (based on study of neutron fluxes and activation), even though the power deposited in the jaws may be higher. This makes the beam of ^{48}Ca radiologically bounding for this device.

A typical distribution of the dose rate on-contact (0 cm) is shown in figure 6. The distribution was obtained with the MARS15 code assuming 30 days of irradiation and 1 day decay after beam shutdown ("30 d/1 d/0 cm" dose rate). Table 3 summarizes both the maximum and average dose rates at various surfaces of the device shielding. Since the shielding size is similar to that of a human "phantom", the average dose rate represents the actual "whole body" exposure more accurately. The average dose rate will stay below 1.7 mrem/h, and the maximum dose rate will be below 8.4 mrem/h.

The top portion of the shielding has a 14 inch diameter hole for the vacuum pump. This hole is substantial and will result in a streaming from the activated jaws. It was estimated that the dose rate above the hole 30 cm above the top shielding will be approximately 50 mrem/h. A barrier around the top device shielding is considered to limit the access.

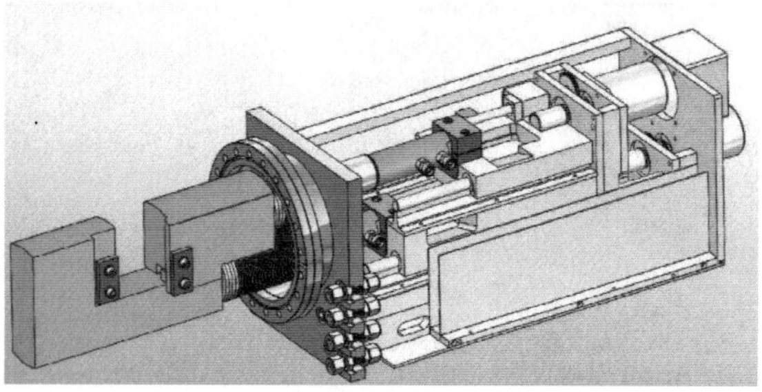

Figure 4. Temporary charge selection device.

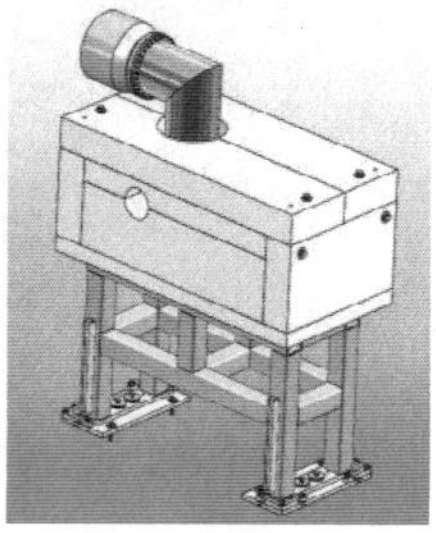

Figure 5. Shielding assembly of the temporary charge selection device.

Table 2. Beam characteristics at the temporary charge selection device.

Beam	Duration (week)	Energy after stripper (MeV/u)	Beam power on stripper (W)	Number of charge states	Fraction of beam to be accelerated (%)	Power loss on charge selection jaws (W)
^{238}U	12	16.5	1250	5	80	250
^{48}Ca	6.34	20	1135	1	90	135
^{78}Kr	2.21	20	1135	3	90	135
^{124}Xe	1.3	17.3	1250	3	80	250
^{18}O	0.86	20	1000	1	100	0
^{86}Kr	0.63	19.3	1130	3	90	130
^{16}O	0.44	20	1000	1	100	0
^{36}Ar		20	1000	1	100	0

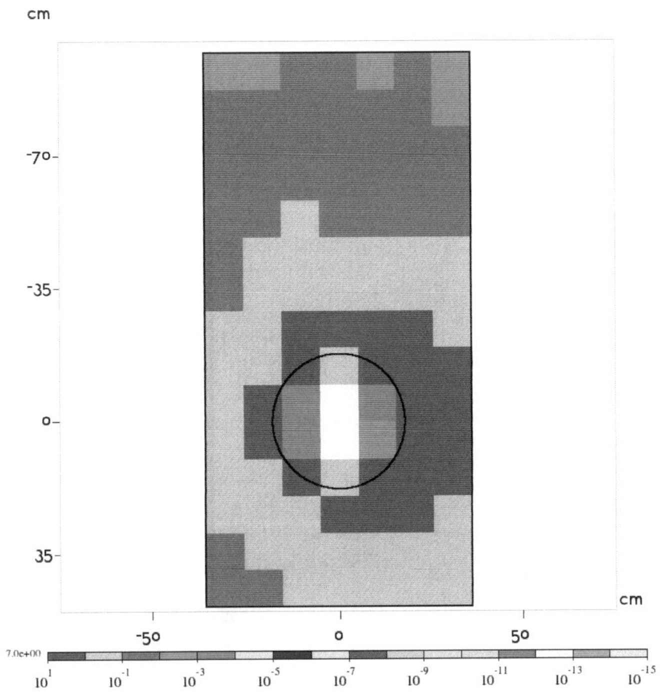

Figure 6. Residual dose rate on-contact at the top shielding of the temporary charge selection device calculated after 30 days of irradiation and 1 day after shutdown. The circle indicates the 14 inch hole for the vacuum pump.

Table 3. Maximum and average residual dose rates on-contact at various shielding surfaces of the temporary charge selection device. The dose rates were calculated assuming 30 days of irradiation and 1 d after shutdown.

Shielding surface	Peak dose rate (mrem/h)	Average dose rate (mrem/h)
Downstream	8.4	1.7
Upstream	0.44	0.15
Top	7.0	0.59
Bottom	1.7	0.41
Left	0.22	0.068
Right	0.72	0.35

5. Beam dumps in Folding Segment 1

There are two small beam dumps in Folding Segment 1 (figure 7). The beam dump names are abbreviated FS1a and FS1b. The beam dump FS1a is intended for tuning up Linac Segment 1. We assumed that the operating duty will be limited to 5% of the operational time of 5556 h a year. The beam power at FS1a is 15 W. The second beam dump, FS1b, is purposed for the study of charge selection. It will be used for only a few days a year but at a higher beam power of up to 500 W. The designs of both dumps and their shielding are similar. Both are made of tungsten with brazed air-cooled copper fins. The dumps are enclosed in steel shielding with the top part (thinnest) to be 4 inches-thick and 6 inches-thick respectively (figure 8 and figure 9). Since the holes in the shielding are small, the direct streaming from the dump cores is relatively small, thus the dose rate at the dumps is driven by the activation of the shielding. The "30 d/1 d/30 cm" dose rate calculated at various

IOP Conf. Series: Journal of Physics: Conf. Series **1046** (2018) 012014 doi:10.1088/1742-6596/1046/1/012014

surfaces of the FS1a assembly for the bounding beam of ^{18}O at 20 MeV/u is insignificant (table 4). For the FS1b dump, the dose rate is higher and reaches 8.12 mrem/h on average for the downstream surface of the shielding assembly (table 5). Radiation levels of bare dump cores are not yet determined. This will be relevant for dump maintenance operations.

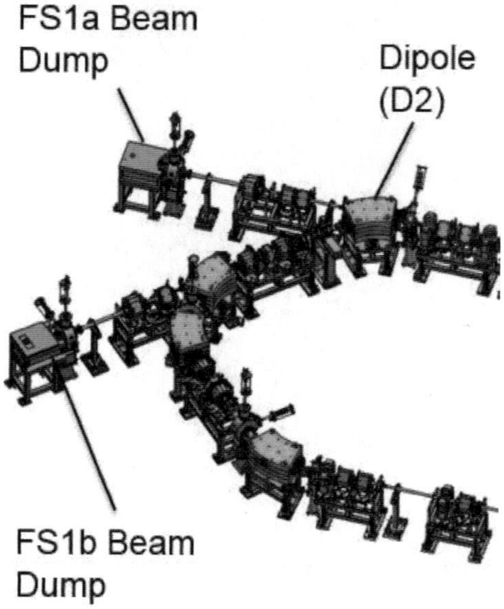

Figure 7. Schematics of Folding Segment 1.

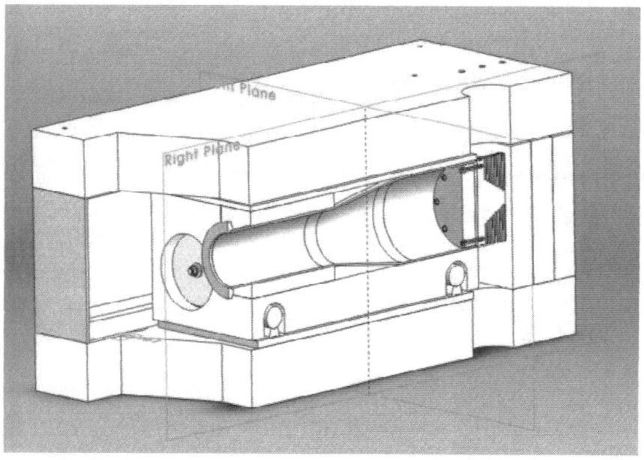

Figure 8. Cut-through view of the FS1a shielding assembly.

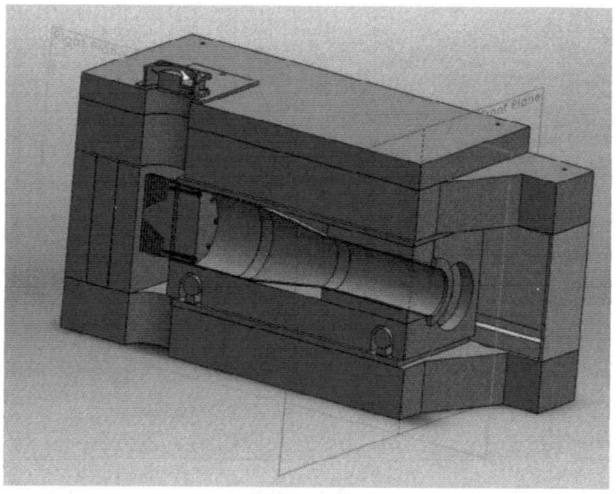

Figure 9. Cut-through view of the FS1b shielding assembly.

Table 4. Maximum and average residual dose rates at various surfaces of the FS1a shielding assembly.

Shielding surface	Maximum dose rate (mrem/h)	Average dose rate (mrem/h)
Top	1	0.09
Bottom	0.5	0.05
Upstream	0.1	0.008
Downstream	1.4	0.17
Side	0.15	0.015

Table 5. Maximum and average residual dose rates at various surfaces of the FS1b shielding assembly.

Shielding surface	Maximum dose rate (mrem/h)	Average dose rate (mrem/h)
Top	9.07	1.81
Bottom	10.2	1.76
Upstream	0.93	0.21
Downstream	29.9	8.12
Side	4.30	0.74

6. Beam dumps in high energy linac segments

The beam dumps in Folding Segment 2 and Beam Delivery Section have similar cores and shielding assemblies (figure 10). The cores are tungsten cylinders with air cooled copper fins. The shielding assemblies around the cores are composed of steel bricks 10 inches in height. An assumption was made that the steel blocks were constructed of A36 steel (type of construction steel). Other types of steel may be used for the construction. We currently consider either cast iron or ductile iron. The steel shielding rests on a concrete slab. The air cooling to the dumps is provided through a channel that runs through the slab and steel assembly (figure 11). The shielding must be sufficient to provide protection against prompt exposure to the workers and public above the ground, residual exposure to the workers in the linac tunnel, and activation of the ground water and soil at the beam dumps. The shielding dimensions were verified against all three radiation sources. Note that different shielding dimensions are driven by different radiation sources. For example, the size of shielding above the dump core was determined by facility design goals to have the prompt dose to workers above the ground level below 0.1 mrem/h and the prompt dose rate to public below 0.001 mrem/h. The shielding below the dump core is validated against the design goals for ground water activation. The size of the assembly in the horizontal plane was estimated to meet the design goals for the residual exposure to workers in the linac tunnel.

Both the dumps are intended for tuning up the corresponding linac section – the FS2 dump will absorb beam passing through Linac Segment 2 while the BDS dump will be used to tune up Linac Segment 3. Although all the beam types can be used, the calculations presented in this section were carried out for the radiologically bounding beam of ^{18}O. Both the dumps were designed to operate at a beam power of up to 135 W and a duty factor of 5%. The beam of ^{18}O will be at 207 MeV/u at the FS2 dump, and at 278 MeV/u at the BDS dump.

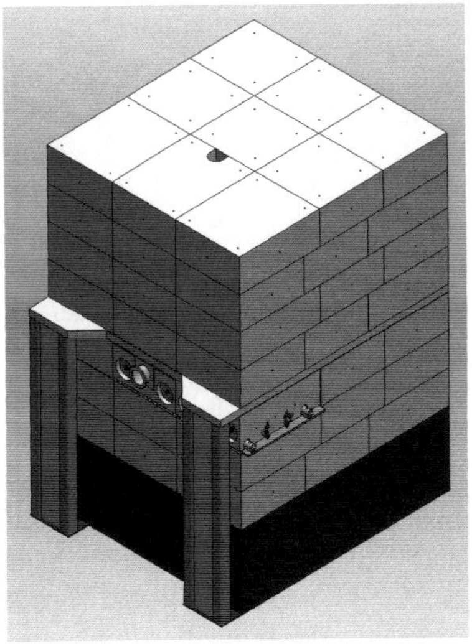

Figure 10. 3-D rendering of the FS2 shielding assembly.

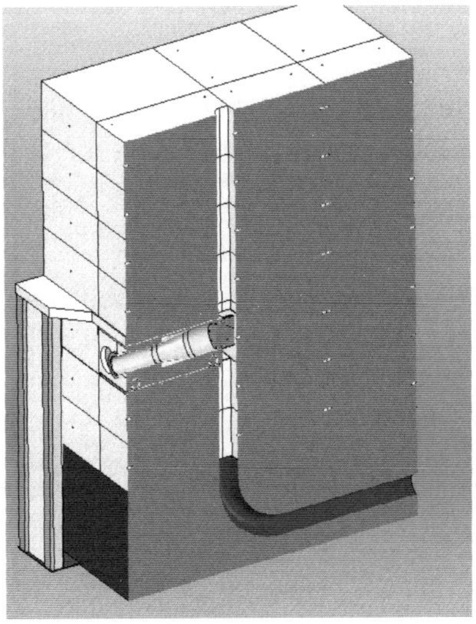

Figure 11. Cut-through view of the FS2 shielding. Air cooling channel is seen.

6.1. Residual activity of beam dump shielding in Folding Segment 2 and Beam Delivery Section

The residual dose on contact for the BDS dump shielding was calculated for 30 days of irradiation and 1 day after beam shutdown. Figure 12 shows a "30 d/1 d/0 cm" distribution in the horizontal plane cutting through the middle of the dump. The isocontours in the distribution indicate that the expected dose rate on the shielding surface will be at a level of a few mrem/h, which meets the design goal. The activation of the FS2 beam dump is expected to be even lower than that due to the lower beam energy.

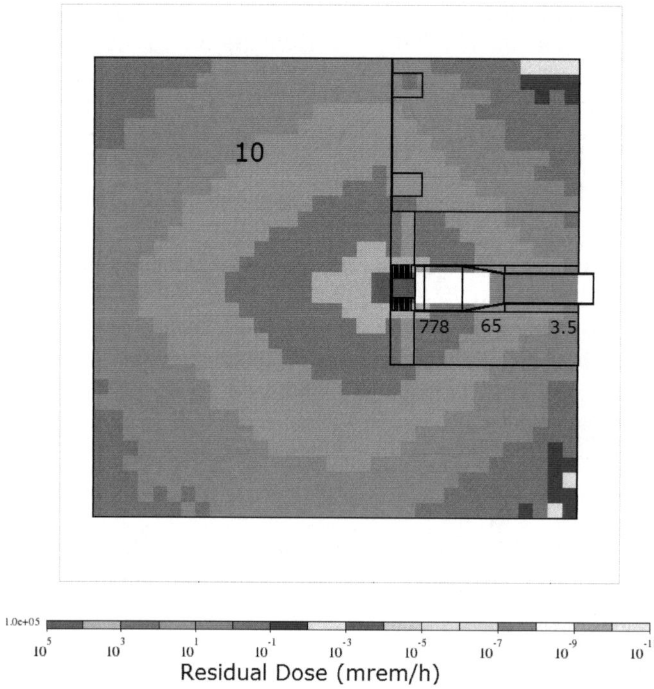

Figure 12. Residual dose rate ("30 d/1 d/30 cm") in the BDS dump shielding at a level of the dump centre.

6.2. Activation of soil and ground water

There is a potential to activate the ground water and soil at the FS2 and BDS beam dumps above the FRIB design requirement to meet the drinking water criteria [8] due to relatively high beam energy. Ground water activation was evaluated based on the previously obtained results for continuous beam losses [9]. As it was found in [9], that the ground water activation approaches the limits for drinking water if the maximum star density immediately outside of the tunnel walls is 10^9 1/cm^3/y. For the beam dumps, the maximum level of activation is below 10^8 1/cm^3/y (figure 13), that is the expected activation level for the ground water should be 10 times below the regulatory limits. The drinking water limits were chosen to provide conservatism in design, even though there are no drinking water sources in the proximity to the facility.

An additional reduction factor is present. As described in [10], the beam dump geometry is typical for a target-like facility. An averaged ground water activation for a target-like facility is approximately 10 times lower than that for continuous losses in a beamline given the maximum star density is the same in both the cases. This was shown in [10] for a simple cylindrical tunnel. In our case, the tunnel geometry is not cylindrical but more complicated, but still some reduction factor is expected. Therefore we can conclude, that the activation of the ground water is expected to be more than 10 times below the regulatory limits for drinking water.

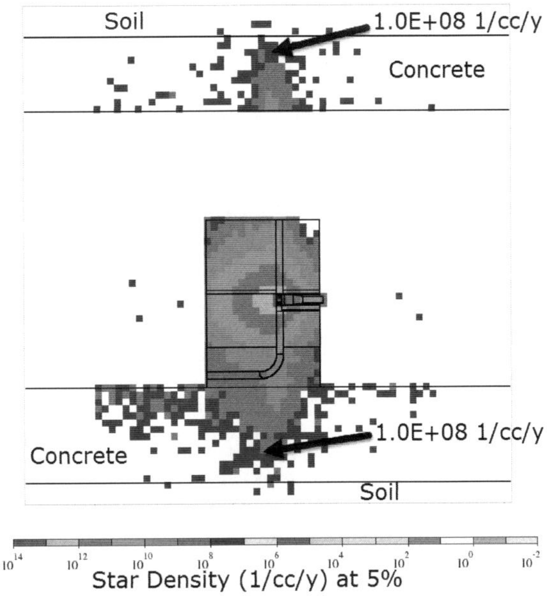

Figure 13. Star density distribution at the BDS beam dump.

6.3. Dose rate from bare dump cores

As figures 10 and 11 showed, the diameter of the penetrations in the dump shielding is relatively small compared to the shielding thickness. Therefore, it is expected that a contribution to the residual dose at the shielding from the dump cores will be relatively small compared to that from the shielding itself. However, the dose rate from the bare dump cores is needed in order to evaluate the worker exposure during the dump maintenance. Figure 14 shows a dose rate distribution at the bare BDS dump core obtained for 720 h of irradiation and 4 h of cooling time. The dose rate at 30 cm from the core reaches 500 mrem/h. An area where the dose rate exceeds 100 mrem/h at 30 cm from the source is designated a "High Radiation Area". The high dose rate will severely limit maintenance activities. Placing an activated dump core in a 5 cm-thick steel casket will reduce the dose down to a more manageable level (figure 15). This calculation serves as a basis for understanding how to replace, handle, store and dispose the dumps.

IOP Conf. Series: Journal of Physics: Conf. Series **1046** (2018) 012014 doi:10.1088/1742-6596/1046/1/012014

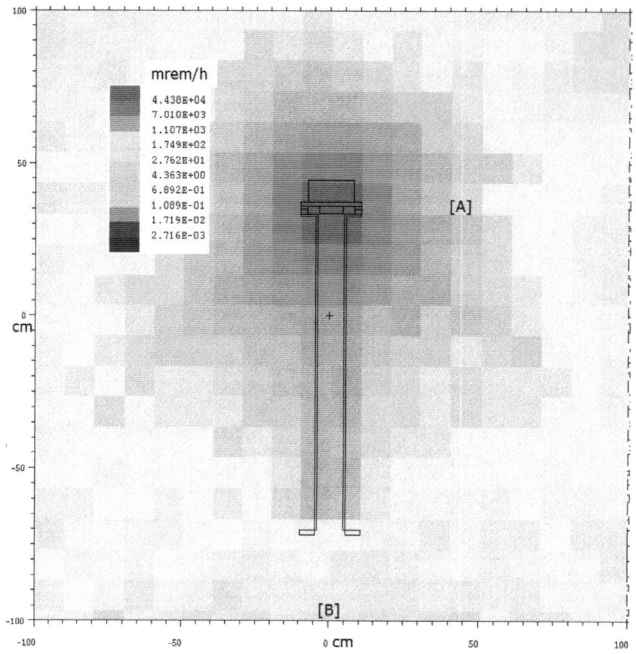

Figure 14. Residual dose rate at bare BDS dump core after 720 h of irradiation and 4 h after shutdown. The dose rate at the location [A] is 498 mrem/h. The dose rate at the location [B] is 37 mrem/h.

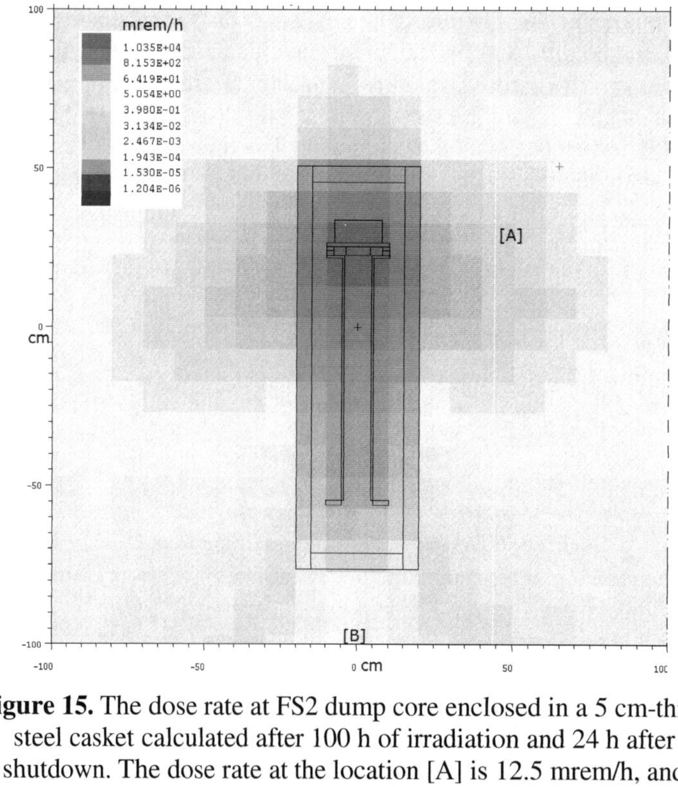

Figure 15. The dose rate at FS2 dump core enclosed in a 5 cm-thick steel casket calculated after 100 h of irradiation and 24 h after shutdown. The dose rate at the location [A] is 12.5 mrem/h, and 13.3 mrem/h at the location [B].

IOP Conf. Series: Journal of Physics: Conf. Series **1046** (2018) 012014 doi:10.1088/1742-6596/1046/1/012014

7. Conclusions

The residual activation of high-rate beam loss devices at the FRIB linear accelerator was evaluated. These devices are the charge stripping device, two beam dumps in Folding Segment 1, beam dump in Folding Segment 2 and beam dump in Beam Delivery Section. The temporary charge selector designed to handle up to 10% of the facility full power was also evaluated. The full power charge selector is not yet designed. The residual activity for all the devices is understood, and their local shielding, if any, was evaluated. No local shielding is planned for the charge stripper. Movable shielding screens will be used where necessary to bring the dose rate to workers from several tens of mrem per hour down to levels that will facilitate operation. The activity of the bare cores of the beam dumps in Folding Segment 2 and Beam Delivery Section can be as high as several hundred mrem per hour depending on the irradiation time, cooldown time and beam type used. Specially designed caskets are considered to shield the bare dump cores for the purpose of maintenance, storage and disposal.

Acknowledgement

This material is based on work supported by the U.S. Department of Energy Office of Science under Cooperative Agreement DE-SC0000661, the State of Michigan and Michigan State University. Michigan State University is designing and establishing FRIB as a DOE Office of Science National User Facility in support of the mission of the Office of Nuclear Physics.

References

[1] United States Nuclear Regulatory Commission *10 CFR 20.1003*

[2] Sihver L, Sato T, Gustafsson K, Mancusi D, Iwase H, Niita K, Nakashima H, Sakamoto Y, Iwamoto Y and Matsuda N 2010 An update about recent developments of the PHITS code *Adv. Space Res.* **45** pp 892-99

[3] Kai T, Maekawa F, Kosako K, Kasugai Y, Takada H, and Ikeda Y 2001 DCHAINSP-2001: high energy particle induced radioactivity calculation code *JAERI-Data/Code 2001-016*

[4] Mokhov N 1995 The MARS code system user's guide *Fermilab-FN-628*

[5] Mokhov N and Striganov S 2007 MARS15 overview *Proc. of Hadronic Shower Simulation Workshop (Batavia Illinois USA) AIP Conf. Proc.* **896** pp 50-60 (*Fermilab-Conf-07/008-AD*)

[6] Mokhov N, Aarnio P, Eidelman Yu, Gudima K, Konobeev A, Pronskikh V, Rakhno I, Striganov S and Tropin I 2014 MARS15 code developments driven by the intensity frontier needs *Prog. Nucl. Sci. Technol.* **4** pp 496-501

[7] https://mars.fnal.gov

[8] United States Environmental Protection Agency national primary drinking water regulations *40 CFR Chapter I Subchapter D Parts 141-143*

[9] Kostin M, Lowrie R and Ronningen R 2015 Evaluation of radiation environment at FRIB linac *Proc. of the Shielding Aspects of Accelerator, Targets and Irradiation Facilities-SATIF-12 (Batavia Illinois USA)* pp 71-87

[10] Cossairt D 2016 Radiation physics for personnel and environmental protection *FNAL-TM-1834*

Activation Levels, Handling, and Storage of Activated Components in the Target Hall at FRIB

D. Georgobiani[1], R. Bennett[1], G. Bollen[1], M. Kostin[1], R. Ronningen[1]

[1] Facility for Rare Isotope Beams (FRIB), Michigan State University, 640 South Shaw Lane, East Lansing, MI 48824 USA

georgobi@frib.msu.edu

Abstract. The Facility for Rare Isotope Beams (FRIB) is a major new scientific user facility under construction in the United States for nuclear science research with beams of rare isotopes. 400 kW beam operations with heavy ions ranging from oxygen to uranium will create a high radiation environment for many components, particularly for the beam line components located in the target hall, where approximately 100 kW of beam power are dissipated in the target and another 300 kW are dissipated in the beam dump. Detailed studies of the component activation, their remote handling, storage, and transport, have been performed to ensure safe operation levels in this environment. Levels of activation are calculated for the beam line components within the FRIB target hall.

1. Radiation Transport Scope within the FRIB Target Facility

The FRIB scope is shown in figure 1. The Target Facility encompasses a pre-target beam delivery system area, production target systems located in the target hall, and the fragment separator. The fragment separator (figure 2) is designed for production and delivery of rare isotopes with high rates and high purities to maximize FRIB science reach. It is planned to accept a primary beam power of 400 kW, with beam energies of more than 200 MeV/u.

The target hall will have a high radiation environment, with two major sources of radiation: beam power dissipated in the target (~100 kW) and beam power dissipated in the beam dump (~300 kW). Both target and beam dump will be periodically replaced. The highest activated component will be post-target magnet shield located downstream of the target module.

Radiation transport analysis supports technical design and safe operation of the entire project. The scope of this paper is the description of some analysis specific to the Target Facility (see figure 1). The radiation transport model of the target hall and its components is based on engineering models developed from mechanical and facility designs that were converted using code MCAM [1] and geometry visualization software VISED [2] for use by the Monte-Carlo codes PHITS [3] and MCNP [4] to evaluate the radiation environment. High level of detail in the model supports construction and preparation for operations.

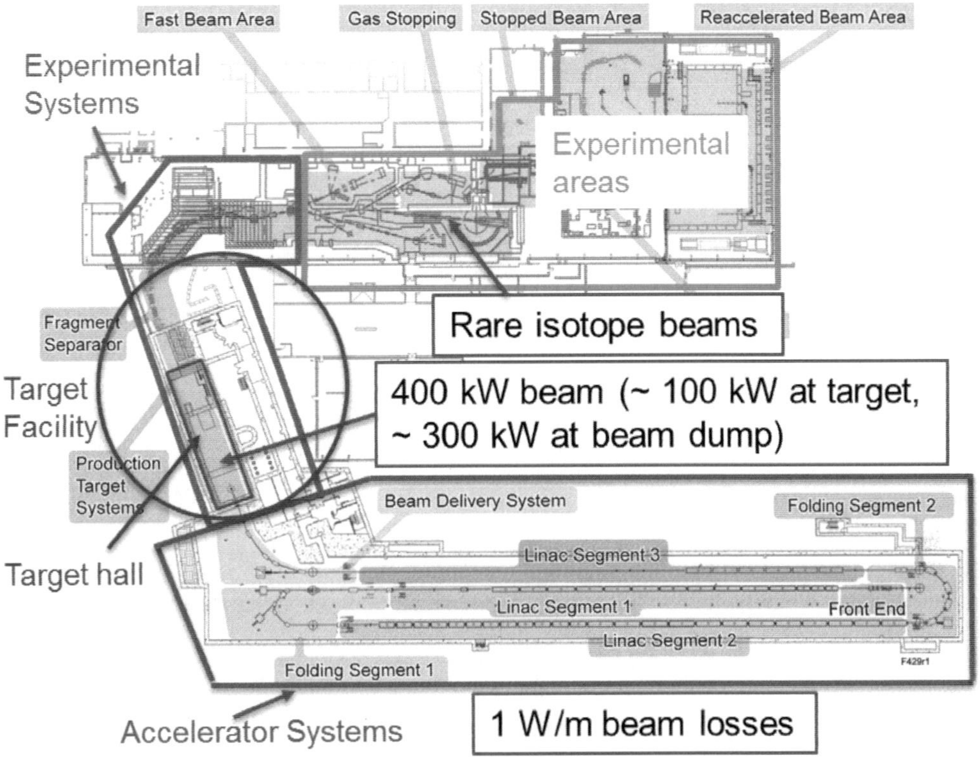

Figure 1. FRIB scope. The Target Facility contains the target hall.

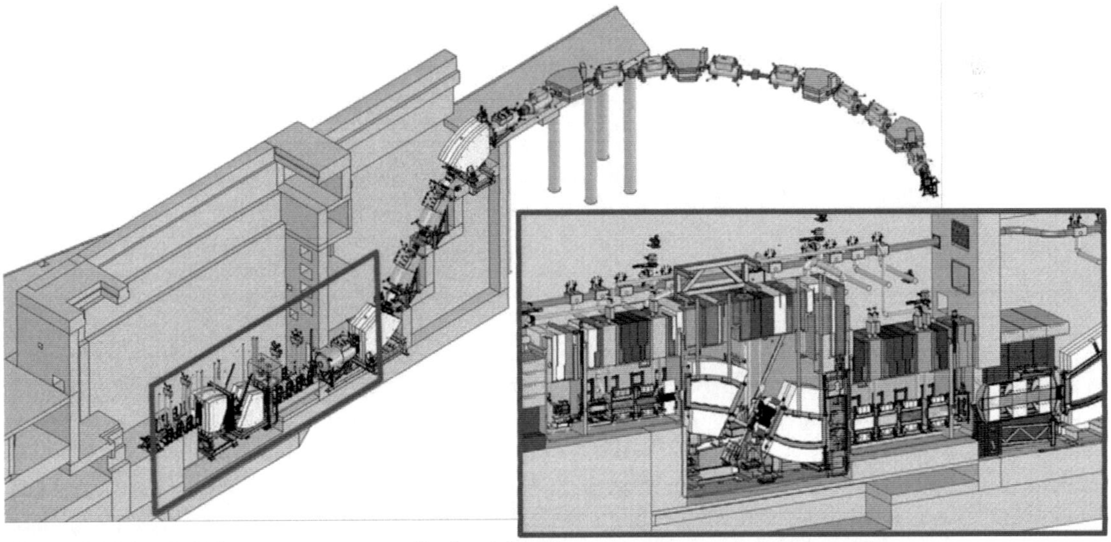

Figure 2. Target hall fragment separator. On the right: Three vacuum vessels with beam line components such as target, beam dump, and magnets; multi-colored, multi-shaped blocks represent re-entrant shielding above the vessels that house the beam line components..

2. Radiation Transport Analysis of Activated Beam Line Components

Activation calculations are performed for all beam line equipment. Conservative beam variety and energy choices are employed. As an example, table 1 shows unshielded dose rates for the highest activated components at 30 cm distance, calculated for a ^{48}Ca beam at 549 MeV/u (a conservative beam/energy assumption), and at 400 kW total power. Irradiation and cooling times are based on planned component service lifetimes and estimated times for preparation activities prior to removal, respectively: targets will be replaced every one to two weeks, depending on experiment, while beam dump will be replaced once a year. The post-target magnet shield, a bronze block with a 40 mrad conical aperture, placed right after the target module to shield the magnets downstream of the target, is planned to last for 30 years, the nominal facility lifetime.

Table 1. Unshielded dose rates from highest activated components at 30 cm.

Component	Irradiation Time	Cooling Time	Dose Rate, rem/h
Target Module	14 days	4 hours	13
Beam Dump Module	1 year	16 hours	51
Post-Target Magnet Shield	30 years	2 days	1547

The high dose rates impact operational scenarios. Highly activated components need to be adequately shielded in order to maintain hands-on access to the target hall when shielding is in place and the beam is off. FRIB minimizes radiation exposure to levels as low as reasonably achievable (ALARA) and well below regulatory limits, and uses engineering and administrative controls to maintain safe radiation exposures in occupied areas. Table 2 lists regulatory limits and MSU ALARA limits for general public and radiation workers applicable to FRIB and used in the present analysis. For reference, 1 mSv = 100 mrem.

Table 2. Radiation protection limits and goals at FRIB.

Radiation Dose – Worker	Standard*: 5,000 mrem/yr MSU ALARA Goal**: 500 mrem/yr	
Radiation Dose – Public	Standard*: 100 mrem/yr and < 2 mrem/(any one hour) MSU ALARA Goal**: 10 mrem/yr and < 2 mrem/(any one hour)	
Air - maximum exposure to nearest receptor	Standard*: 10 mrem/y MSU ALARA Goal**: 1 mrem/yr	
Groundwater – effluent	^{3}H Standard*: 1,000 pCi/ml FRIB Design Goal: 20 pCi/ml (drinking water standard***)	^{22}Na Standard*: 6 pCi/ml FRIB Design Goal: 0.4 pCi/ml (drinking water standard***)

3. Remote Handling

A remote handling system is used to handle highly activated components. The systems consists of a remotely operated bridge 20 ton crane, a window workstation with master-slave-manipulators, a remote viewing system (cameras), a remote handling equipment lift, and tooling for handling.. When the beam is off and re-entrant shielding (see figure 2) of the vacuum vessels is in place, hands-on access to the target hall is possible. The estimated 80 hours/year time allowed for personnel access to the target hall for target changes results in personnel doses below MSU ALARA limit of 500 mrem/year. Operations also can include manual utility disconnect, re-entrant lid bolting, as well as general cell maintenance. Whenever the beam is on, or the re-entrant shielding is removed, operations are remote and use the crane and/or window workstations, and include shielding removal, component removal, and waste handling. The window workstation is used for all dexterous operations, such as component maintenance.

3.1. Dose rates to personnel

Activated components in the target hall will be brought to the Master Slave Manipulator (MSM) gallery's lead glass window for inspection and repair. The MSM gallery is located on the other ("cold") side of a target hall wall, and working personnel will be present there. Dose rates in the MSM gallery due to activated components help decide whether adequate measures, such as better local shielding or limited access, should be implemented to protect working personnel from radiation exposure. Table 3 shows dose rates in the MSM gallery, 30 cm from the window, resulting from highest-activated components. The components are assumed to be placed 1 m away from the lead glass window on the target hall ("hot") side. The size of the lead glass is 70 cm x 56 cm, and the thickness is designed to provide radiation protection equivalent to that of the surrounding concrete wall. Irradiation time of 30 years (lifetime of the facility) is assumed. Resulting dose rates are compared to 0.1 mrem/h (this number is the FRIB shielding design goal and it comes from table 2, Radiation Dose to Workers, assuming 5000 operational hours per year). Dose rates are higher than 0.1 mrem/h for the post-target magnet shield; but the shield is expected to last for lifetime of facility. Dose rates are ~ 0.1 mrem/h for the beam dump module, and < 0.1 mrem/h for the target module. The assumptions are conservative; dose rates for the same irradiation and cooling times are calculated to be smaller for heavier beams.

Table 3. Dose rates in the MSM gallery from highest activated components.

Beam/energy	**48-Ca, 240 MeV/u**		**48-Ca, 549 MeV/u**		
Component	**Magnet Shield**	**Target Module**	**Magnet Shield**	**Target Module**	**Beam Dump**
Cooling time	**Dose rate, mrem/h**				
4 hours	0.79	1.5e-2	3.2	4.2e-2	
16 hours					0.125
24 hours	0.71	1.2e-4	3.0	3.2e-4	
5 days	0.67	3.3e-5	2.9	9.3e-5	

3.2. Waste Storage

The beam line components have modular designs. When a module fails, it is removed from the component, packed in a "waste basket", and placed in a canyon (area next to the target hall vacuum vessels) for temporary storage. When enough failed modules are accumulated, shipment to a disposal site is scheduled. The failed modules are then removed from the canyon, and packed in a shipping container. The shipping container is then placed in a cask for shipment.

3.2.1. Target hall canyon shielding

Since activated components and waste, such as target assemblies, wedge assemblies, target disk modules, as well as scrap cryostats, are planned to be temporarily stored in target hall canyon, the canyon radiation environment was assessed for a conservative beam/energy combination: ^{48}Ca beam at 549 MeV/u. Irradiation time of 30 years and cooling time of 4 hours were assumed. Resulting dose maps show that the canyon needs to be shielded when personnel are present (figure 3). The canyon cover shield thickness was optimized so that dose equivalent rates above the cover are below 5 mrem/h (dose rate of 5 mrem/h and higher to 100 mrem/h is the definition of a radiation area). This is at or below the level of activation expected from general target hall activation. If the post-target magnet shield (life-of-facility component) is not present in the canyon, then a 2-ft thick concrete cover is needed. If the magnet shield is stored in the canyon, 2-in thick local steel shield is needed in addition to the 2-ft thick concrete cover to keep the dose equivalent rates outside of the canyon below 5 mrem/h.

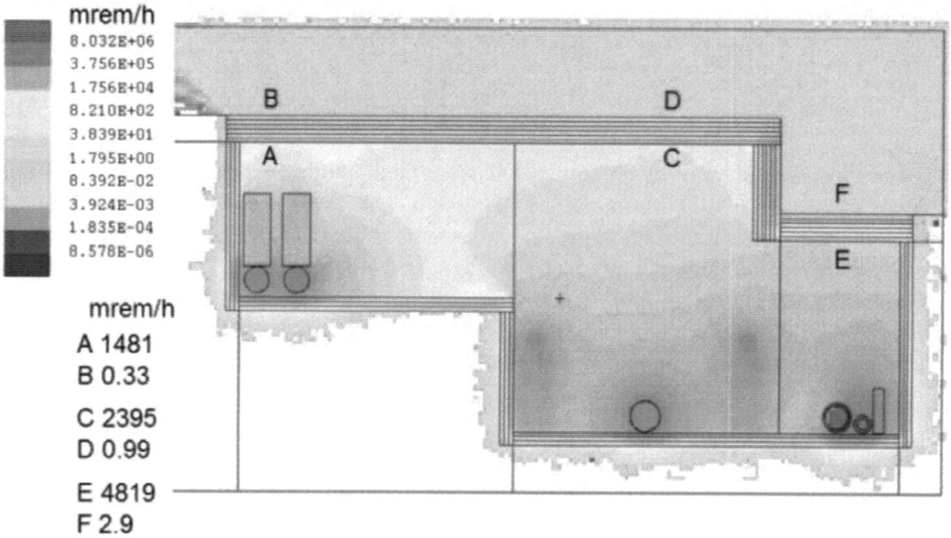

Figure 3. Dose equivalent rates due to activated components placed in the target hall canyon, an area next to the vacuum vessels.

4. Conclusions

FRIB is being designed and established at MSU as a national user facility to provide fast, stopped, and reaccelerated beams of rare isotopes. High radiation environment in the project target hall demands analysis of beam line component activation. Highly activated components that need to be fixed or disposed will be temporarily stored for cooling and/or disposal. Levels of activation and dose rates have been studied to minimize personnel doses during handling and storage.

Acknowledgments

This material is based on work supported by the U.S. Department of Energy Office of Science under Cooperative Agreement DE-SC0000661, the State of Michigan and Michigan State University. Michigan State University designs and establishes FRIB as a DOE Office of Science National User Facility in support of the mission of the Office of Nuclear Physics.

References

[1] Y. Wu, FDS Team, CAD-based interface programs for fusion neutron transport simulation, Fusion Engineering and Design 84 (2009) 1987-1992

[2] http://www.mcnpvised.com/

[3] T. Sato, K. Niita, N. Matsuda, S. Hashimoto, Y. Iwamoto, S. Noda, T. Ogawa, H. Iwase, H. Nakashima, T. Fukahori, K. Okumura, T. Kai, S. Chiba, T. Furuta and L. Sihver, Particle and Heavy Ion Transport Code System PHITS, Version 2.52, J. Nucl. Sci. Technol. 50:9, 913-923 (2013)

[4] D. B. Pelowitz, ed., MCNPX User's Manual, Version 2.7.0, Los Alamos National Laboratory report LA-CP-11-00438 (2011)

IOP Conf. Series: Journal of Physics: Conf. Series **1046** (2018) 012016 doi:10.1088/1742-6596/1046/1/012016

In-situ determination of residual specific activity in activated concrete walls of a PET-cyclotron room

H Matsumura[1], A Toyoda[1], K Masumoto[1], G Yoshida[1], T Yagishita[2], T Nakabayashi[2], H Sasaki[2], K Matsumura[2], Y Yamaya[2] and Y Miyazaki[3]

[1]Radiation Science Center, High Energy Accelerator Research Organization (KEK) 1-1 Oho, Tsukuba, Ibaraki 305-0801, Japan
[2]Japan Environment Research Co., Ltd., 6-24-1 Nishi-Shinjuku, Shinjuku-ku, Tokyo 160-0023, Japan
[3]The Medical and Pharmacological Research Center Foundation, Wo-32, Inoyama-machi, Hakui, Ishikawa 925-0613, Japan

hiroshi.matsumura@kek.jp

Abstract. In the decommissioning work for concrete walls of PET-cyclotron rooms, an in-situ measurement is expected to be useful for obtaining a contour map of the specific activity on the walls without destroying the structure. In this study, specific activities of γ-ray-emitting radionuclides in concrete walls were determined by using an in-situ measurement method employing a portable Ge semiconductor detector, and compared with the specific activity obtained using the sampling measurement method, at the Medical and Pharmacological Research Center Foundation in Hakui, Ishikawa, Japan. Accordingly, the specific activity could be determined by the in-situ determination method. Since there is a clear correlation between the total specific activity of γ-ray-emitting radionuclides and contact dose rate, the specific activity can be determined approximately by contact dose-rate measurement using a NaI scintillation survey meter. The specific activity of each γ-ray-emitting radionuclide can also be estimated from the contact dose rate using a NaI scintillation survey meter. The in-situ measurement method is a powerful tool for the decommissioning of the PET cyclotron room.

1. Introduction

Radioactive isotopes, such as ^{18}F, ^{15}O, ^{13}N, and ^{11}C, are produced as tracers for positron emission tomography (PET). Proton cyclotrons with acceleration energies of between ten and twenty MeV are used to produce these radioactive isotopes. There are more than 140 cyclotrons for medical use in Japan, and it is estimated that many of them will be decommissioned due to aging in the near future. Therefore, the decommissioning methodology for these accelerator facilities is an important issue.

Radioactive isotopes for PET are produced by proton-induced nuclear reactions in production targets. Neutrons are also produced simultaneously in the nuclear reactions. The produced neutrons flight into the accelerator body and the concrete walls of the accelerator room. Consequently, the neutrons induce nuclear reactions in these materials, and activate them. For the decommissioning of the accelerator facility, "radioactive materials" must be isolated. The concrete walls in the accelerator room are especially huge. The establishment of a determination method for residual specific activity in the concrete walls is important for the efficient decommissioning of accelerator facilities.

IOP Conf. Series: Journal of Physics: Conf. Series **1046** (2018) 012016 doi:10.1088/1742-6596/1046/1/012016

Before the decommissioning of the accelerator facility, the residual specific activity in the concrete walls has to be determined to isolate the radioactive material region. At present, the determination of specific activity in concrete is carried out by using γ-ray spectrometry with a Ge semiconductor detector (Ge detector) after sampling by boring the concrete core [1, 2]. In this "sampling measurement method", the depth variations of the residual specific activities of radionuclides can be determined.

However, the sampling measurement method has the following two drawbacks. (1) Specific skills for radiological work are required for the concrete sampling. (2) Since the sampling number is limited, it is difficult to map for identifying the whole image of activation. If the residual specific activity in the concrete wall is determined non-destructively by attaching a detector to the concrete wall (in-situ measurement method), a contour map of the specific activity can be obtained in detail without destroying the structure.

The establishment of the in-situ measurement method can improve the efficiency of the decommissioning process and reduce radioactive waste and disposal cost as much as possible. Therefore, in this study, we determined the residual specific activity in concrete walls by using a portable Ge detector and a NaI scintillation survey meter (NaI survey meter) which were brought into an actual PET cyclotron room during decommissioning. The sampling measurement was also carried out, and the validity of the determination results was evaluated by a comparison between the in-situ measurement method and the sampling measurement method.

2. Experimental procedures

2.1. Investigation of a PET-cyclotron facility

An NKK-Oxford superconducting cyclotron [3] used for the production of radiopharmaceuticals for PET at the Medical and Pharmacological Research Center Foundation in Hakui, Ishikawa, Japan was decommissioned in March 2015. The concrete walls of this PET cyclotron room were investigated in this study. Information on the PET-cyclotron facility is given in Table 1.

The energy and output current of the proton beam are 12 MeV and 60 μA, respectively. Since the cyclotron was used for the production of PET radiopharmaceuticals for delivery to hospitals, the operation time per day of the cyclotron (~7 h) was much longer than that of general production use at a hospital (~1 hour or less). Consequently, the activation level in the cyclotron facility was estimated to be eight times higher than that in general hospital cyclotron facilities, considering the operation time of the cyclotron and neutron flux in the room.

The operation time of the PET-cyclotron every year between 1998 and 2014 is shown in figure 1. Although the operation period is from 1998 to 2015, most of the 17618 hours of operation time occupies the ten years from 2005–2014. The decommissioning work began immediately in 2015, and the entire cyclotron body and its accessories were removed from the cyclotron room. Currently, only the concrete walls of the cyclotron room remain unchanged.

Table 1. Information on the PET-cyclotron facility investigated in this study.

Accelerator	Superconducting cyclotron (NKK/Oxford)
Beam	Particle: protons, Energy: 12 MeV, Output current: 60 μA
Purpose of use	Production of PET radiopharmaceuticals for delivery to hospitals
Operation start	July, 2000 (commissioning in 1998 and 1999)
Decommissioning	March, 2015
Integral operation time	17618 h
Room size	5.6 m (north-south) and 5.2 m (east-west)
Height of a ceiling	2.7 m

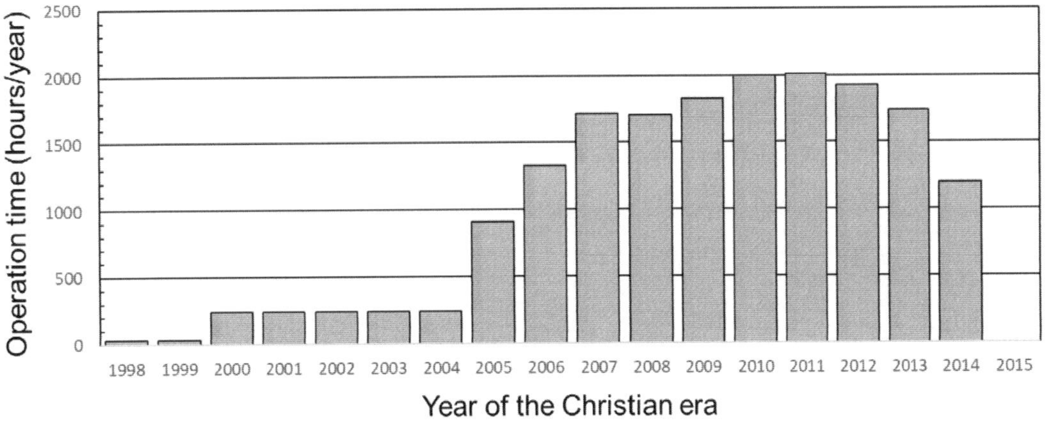

Figure 1. Operation time of the PET-cyclotron each year at the Medical and Pharmacological Research Center Foundation in Hakui, Ishikawa, Japan.

2.2. Sampling measurement method using Ge detector

The specific activities of γ-emitting radionuclides in the concrete walls were determined after concrete core boring. The locations where the concrete core boring was performed are shown as CW1, CW2, CW4, and CW5 in figure 2. The details of this method are described below.

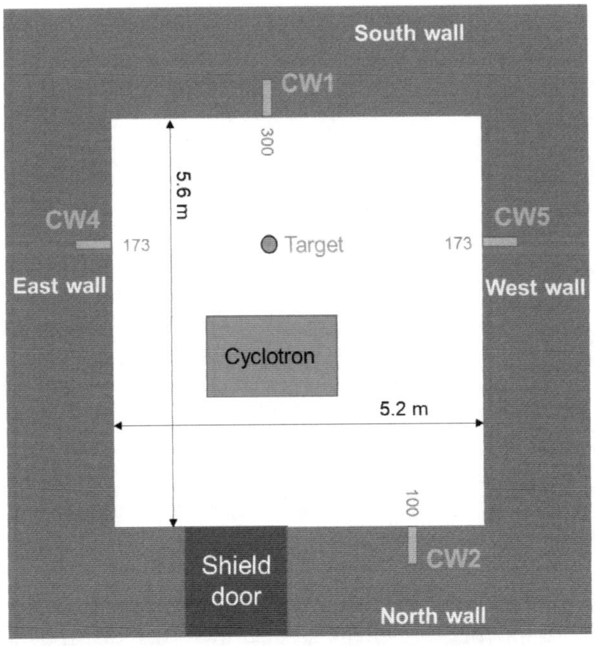

Figure 2. Core boring locations of CW1, CW2, CW4, and CW5. The height of the core boring is 131 cm from the floor. The red values indicate the distances of the CW4 and CW5 cores from the south wall and the distances of the CW1 and CW2 cores from the west wall in cm unit.

First, a 1.5-cm-thick gypsum board was cut and removed from the surface of the concrete walls. Second, 6.5-cm-diameter concrete cores (length: ~50 cm) were collected from the concrete walls. Third, the cores were sliced into 2-cm-thick disks and then milled with a stamp mill. Finally, the powdered concrete samples were transferred into U8 plastic containers and γ-ray spectrometry was performed using Ge detectors to determine the specific activities.

The γ-ray energies used for analysis were 344 keV for ^{152}Eu, 1333 keV for ^{60}Co, 605 keV for ^{134}Cs, 835 keV for ^{54}Mn, and 1461 keV for ^{40}K. The detection efficiencies of the Ge detectors were determined using a standard γ-ray source mixed ^{57}Co, ^{60}Co, ^{85}Sr, ^{88}Y, ^{109}Cd, ^{113}Sn, ^{137}Cs, ^{139}Ce, and ^{203}Hg having the same shape as the samples. The nuclear data were taken from [4]. The specific activities of ^{152}Eu, ^{60}Co, ^{134}Cs, and ^{54}Mn in January 2017 could be obtained.

2.3. Specific activity measurement of a fraction of the reinforcing rod
A fraction of the reinforcing rod (diameter: 1.9 cm, length: ~5.5 cm) was found in the CW1 concrete core. The specific activity of ^{60}Co in the isolated reinforcing rod was determined by γ-ray spectrometry using a Ge detector. The Ge detector was characterized for the efficiency calculation software, ISOCS, by Canberra Inc. [5, 6]. The detection efficiencies of a ^{60}Co 1333-keV γ-ray from the fraction of the reinforcing rod were obtained by using ISOCS. The nuclear data were taken from [4].

2.4. In-situ measurement method using a Ge detector
For in-situ measurements, a portable Ge detector was kept in contact with the surface of the concrete walls. The in-situ measurements were performed at twenty locations in January 2017. They are shown in figure 3. To limit the detection to close range from the Ge detector, the detector was shielded from ambient γ rays by a 6.5-cm-thick and 22.0-cm-long cylindrical Pb shield with a hole of 9.0-cm diameter. Measurements of γ rays were typically performed for ~1500 s (1399–57792 s). The γ-ray energies used for the analysis were 1408 keV for ^{152}Eu, 1333 keV for ^{60}Co, 605 keV for ^{134}Cs, 723 keV for ^{154}Eu, 835 keV for ^{54}Mn, and 1461 keV for ^{40}K. The nuclear data were taken from [4].

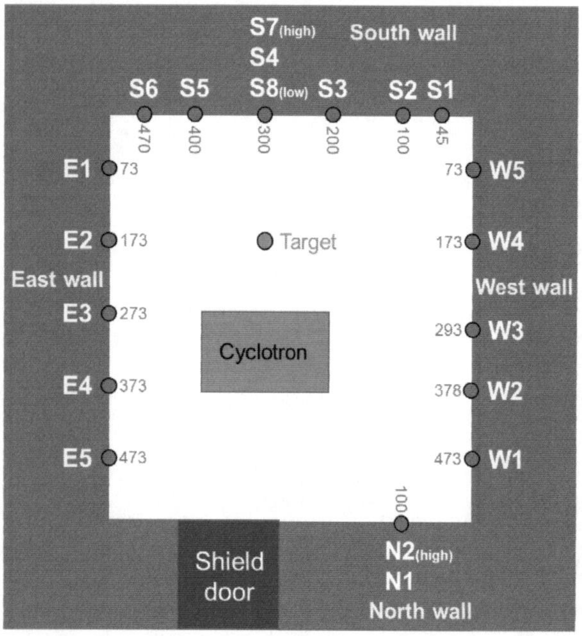

Figure 3. Twenty locations where the in-situ measurement was performed using the Ge detector (red circles). Basically, the height of the measurement was 91 cm from the floor. The heights of S8, S7, and N2 are 51, 151, and 151 cm from the floor, respectively. The red values indicate the distances of the locations of W1-W5 and E1-E5 from the south wall and the distances of the locations of S1-S8, N1, and N2 from the west wall in cm unit.

The portable Ge detector was characterized for ISOCS. The detection efficiency for γ rays from the concrete walls was obtained by using ISOCS, assuming that the γ-ray source was distributed uniformly in the concrete wall. The measured density (2.30 g/cm^3) and analysed chemical composition were input for the efficiency calculation. In the uniform activity condition, the attainable region of γ rays, where the count contribution occupied over 90%, depend on the γ-ray energy. The attainable regions were estimated to have diameter and depth of approximately 40 cm and 17 cm for ^{152}Eu, ^{60}Co, and ^{40}K, respectively, and 30 cm and 13 cm for ^{134}Cs, ^{154}Eu, and ^{54}Mn, respectively, by using ISOCS. The specific activities of ^{152}Eu, ^{60}Co, ^{134}Cs, ^{154}Eu, and ^{54}Mn in the concrete walls were determined.

2.5. *In-situ contact dose rate measurement using a NaI survey meter*

In January 2017 (at the same time when the in-situ measurement using Ge detector was carried out), a NaI survey meter was also attached to the surface of the concrete walls, to obtain a correlation curve between the specific activity and contact dose rate. The measurements of the contact dose rate were performed at the same locations as those of the Ge detector measurements, as shown in figure 3. The model of the NaI survey meter used for this measurement was TCS-171 (Hitachi Aloka Medical, Ltd.). The NaI survey meter was shielded by the same Pb shield as the Ge detector described in Section 2.4. The time constant was set to 10 s. The measurement at a location was performed for a time longer than 30 s. The contact dose rate from a natural background, such as ^{40}K, U series, and Th series in concrete, was obtained as 0.07 μSv/h by measuring a concrete wall in a non-activated adjoining room.

3. Results and discussion

3.1. *Depth profiles of the specific activities*

By the sampling measurement method using Ge detectors described in Section 2.2, the depth profiles of the specific activities of ^{152}Eu, ^{60}Co, ^{134}Cs, and ^{54}Mn in the concrete walls were obtained. The depth profiles of the specific activities from the CW1, CW2, CW4, and CW5 concrete core samples have been plotted in figure 4. The specific activities in decreasing order were ^{40}K, ^{152}Eu, ^{60}Co, ^{134}Cs, and ^{54}Mn. Especially, the specific activities of ^{152}Eu and ^{60}Co account for a substantial fraction of the total specific activity, except for natural ^{40}K. Both ^{152}Eu and ^{60}Co are important radionuclides for the decommissioning of the concrete walls.

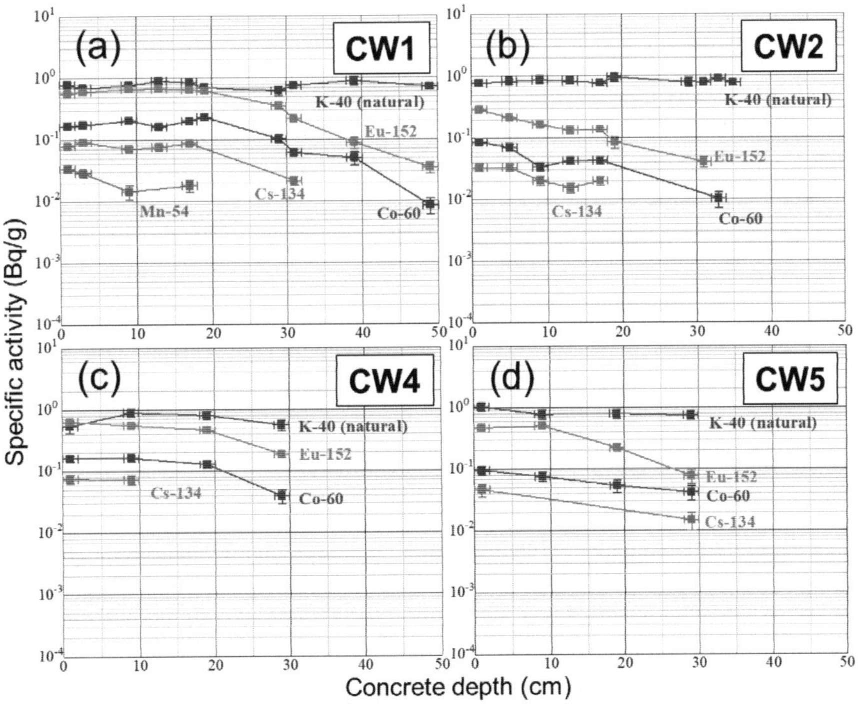

Figure 4. Depth profiles of the specific activities of ^{152}Eu, ^{60}Co, ^{134}Cs, ^{54}Mn, and ^{40}K in the concrete walls from (a) CW1, (b) CW2, (c) CW4, and (d) CW5 concrete core samples in January 2017.

The depth profiles were similar among the radionuclides. Furthermore, the depth profiles were similar among the four different walls. Typically, the specific activity was approximately constant from the surface to a depth of ~20 cm. Thereafter, the specific activity was found to decrease exponentially. In the in-situ measurement, most of the counting comes from the depth shallower than 17 cm, as described in Section 2.4. The attainable region of the γ rays agreed with the constant region of the specific activity. Therefore, the distribution of the γ-ray source in the concrete walls could be considered uniform for the in-situ measurement. This means that the specific activities could be determined approximately by the in-situ measurement method using the Ge detector.

3.2. Spatial distributions of the specific activities

In Section 3.1, it was found that the specific activities could be determined approximately by the in-situ measurement method. Figure 5 shows the spatial distributions of the specific activities of ^{152}Eu, ^{60}Co, ^{134}Cs, ^{154}Eu, ^{54}Mn, and ^{40}K in the concrete walls, which were determined by the in-situ measurement method using the Ge detector, as described in Section 2.4.

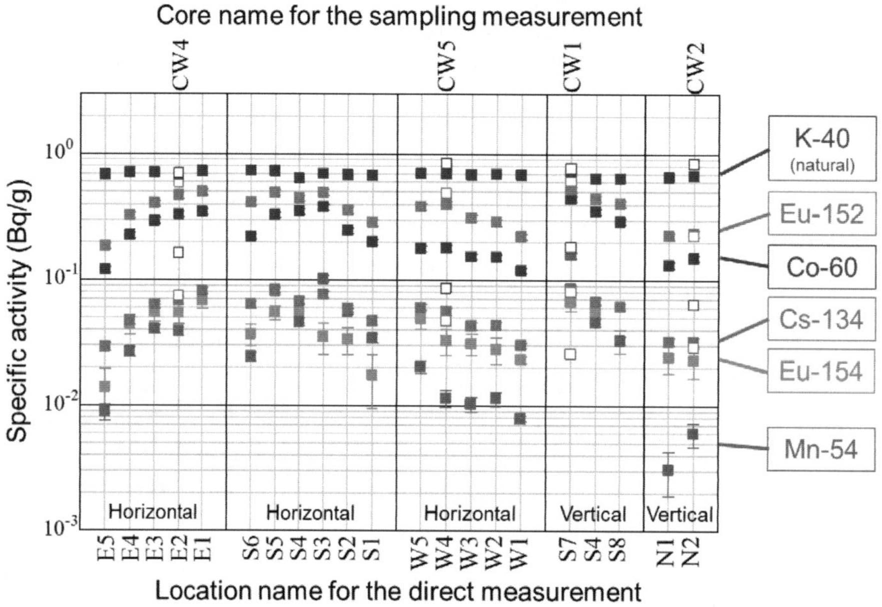

Figure 5. Spatial distributions of the specific activities of ^{152}Eu, ^{60}Co, ^{134}Cs, ^{154}Eu, ^{54}Mn, and ^{40}K in the concrete walls in January 2017. The closed squares indicate the specific activities determined by the in-situ measurement method using the Ge detector. The open squares indicate averages of the specific activities from 0 to 10 cm depth determined by the sampling measurement method.

The distribution of the ^{40}K specific activity was constant because of the inclusion of ^{40}K naturally in the concrete walls. The specific activity distributions were similar among ^{152}Eu, ^{60}Co, ^{134}Cs, and ^{154}Eu. The thermal neutron capture reaction is common in these radionuclide productions. Near the target, the activation level was high. The specific activity decreased with increasing distance from the target. The distribution of the ^{54}Mn specific activity was different from that of the others. In the concrete walls, ^{54}Mn would be produced by ^{54}Fe(n, p)^{54}Mn. Since the specific activity of ^{54}Mn is much smaller than that of the others, ^{54}Mn is not an important radionuclide for the decommissioning of the concrete walls.

In figure 5, the specific activities determined by the in-situ measurement method (closed squares) are compared with those determined by the sampling measurement method (open squares). The specific activities of ^{152}Eu, ^{134}Cs, ^{154}Eu, and ^{40}K determined by the in-situ measurement method agreed well with those obtained by the sampling measurement method.

However, in every concrete wall, the specific activity of ^{60}Co determined by the in-situ measurement method was higher than that in the sampling measurement method. The ratios of the ^{60}Co specific activities (sampling/in-situ) were approximately constant (average: 0.45) regardless of the location. In this facility, 10-cm mesh grids of reinforcing rods with 1.9-cm diameter were placed at a depth of 5 cm from the concrete surface. The reinforcing rods as well as the concrete walls were exposed to thermal neutrons. In the reinforcing rods, ^{60}Co could be produced by the thermal neutron capture reaction of the ^{59}Co impurity. A fraction of the reinforcing rod was obtained from the CW1 concrete core, as described in Section 2.3. The specific activity of ^{60}Co in the isolated reinforcing rods was determined as 3.04±0.15 Bq/g in January 2017.

Next, the count rate of 1333-keV γ-rays from the ^{60}Co of reinforcing rods in the in-situ measurement method was estimated using the measured specific activity in the reinforcing rod and counting efficiency calculated by ISOCS. Accordingly, the count rate ratio of the ^{60}Co γ rays ([estimated from the reinforcing rod]/[measured]) was estimated to be 0.58, which corresponded to the ratios of 0.42 of the ^{60}Co specific activities ([sampling measurement]/[in-situ measurement]). This value agreed well with the measured ratio of 0.45. Therefore, the discrepancy in the ^{60}Co specific activity in the concrete walls between the sampling measurement method and in-situ measurement method was attributed to the contribution from the reinforcing rods. In the in-situ measurement method, the ^{60}Co specific activities in the concrete walls were obtained by multiplying 0.45 as a correction factor.

3.3. Specific activity ratios of each radionuclide for ^{152}Eu

The specific activity ratios of each radionuclide for ^{152}Eu in the concrete walls determined by the in-situ measurement method are plotted in figure 6. The specific activity ratios were independent of the location (except for ^{54}Mn). This fact indicated that if the specific activity of ^{152}Eu is known in some way, the specific activities of the others could be estimated by using the specific activity ratios. The specific activity ratios were determined as 0.24 for ^{60}Co/^{152}Eu, 0.15 for ^{134}Cs/^{152}Eu, 0.105 for ^{154}Eu/^{152}Eu, and 0.096 for ^{54}Mn/^{152}Eu.

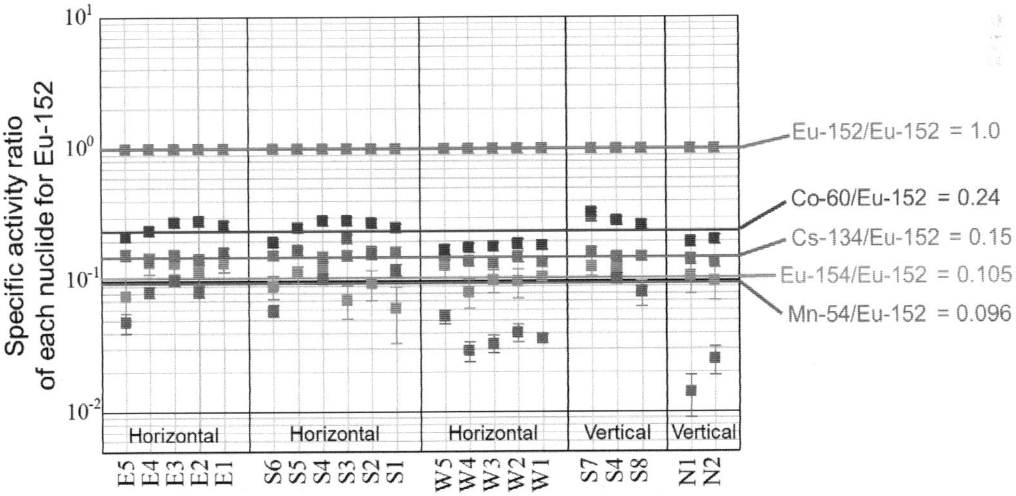

Figure 6. Specific activity ratios of each radionuclide for ^{152}Eu in the concrete walls. The lines indicate the averages of the ratios. The ^{60}Co specific activities corrected for the interference from the reinforcing rods were used for calculating the specific activity ratios of ^{60}Co/^{152}Eu.

3.4. Correlation between the total specific activity and net contact dose rate

The in-situ measurements of contact dose rates were carried out using the NaI survey meter as described in Section 2.5. A correlation curve between the net contact dose rate and the total specific activity in the concrete walls is shown in figure 7. The net contact dose rate means a dose rate obtained by subtracting the natural background described in Section 2.5. The total specific activity was the sum of the specific activities of ^{152}Eu, ^{60}Co, ^{134}Cs, ^{154}Eu, and ^{54}Mn, which were determined by the in-situ measurement method using the Ge detector. There is a clear correlation between the total specific activity and net contact dose rate. Therefore, the net contact dose rate on the concrete wall (D) can be used for estimating the total specific activity in the concrete wall (A_{total}), as A_{total} (Bq/g) = D (μSv/h) ×2.1.

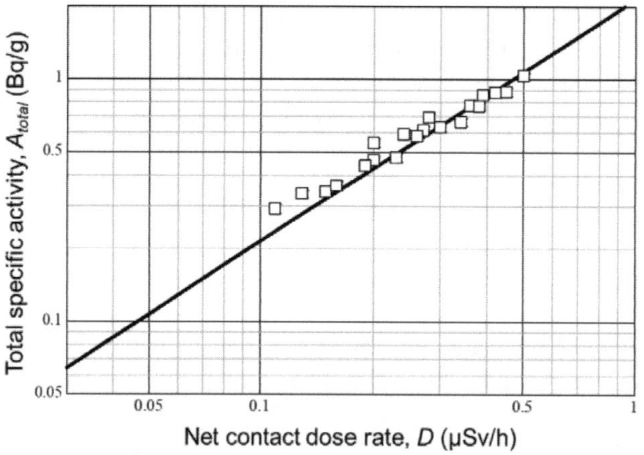

Figure 7. Correlation curve between the net contact dose rate and the total specific activity in the concrete walls.

3.5. Estimation of specific activities of the radionuclides from the net contact dose rates

A correlation curve between the net contact dose rate and specific activity of ^{152}Eu in the concrete walls is shown in Figure 8. There is also a clear correlation between the ^{152}Eu specific activity ($A_{Eu\text{-}152}$) and the net contact dose rate (D), as $A_{Eu\text{-}152}$ (Bq/g) = D (μSv/h) ×1.3.

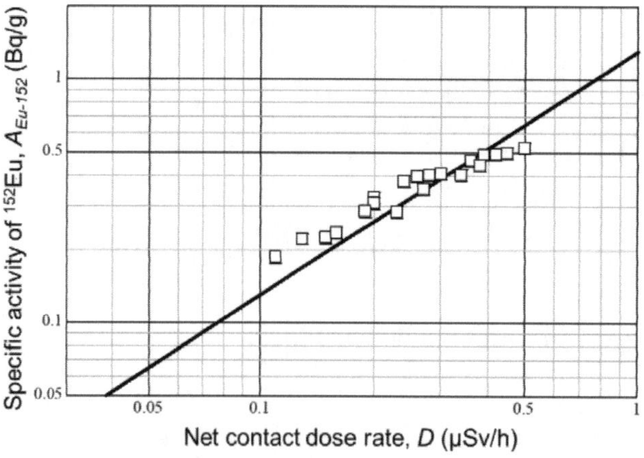

Figure 8. Correlation curve between the net contact dose rate and specific activity of ^{152}Eu in the concrete walls.

In Section 3.3, it was found that the specific activity ratios of each radionuclide for ^{152}Eu in the concrete walls were approximately constant, regardless of the location. Therefore, the specific activities of ^{60}Co, ^{134}Cs, ^{154}Eu, and ^{54}Mn can be estimated from the net contact dose rate and the specific activity ratio of each radionuclide for ^{152}Eu. In Figure 9, the specific activities estimated from the net contact dose rate and the specific activity ratio of each radionuclide for ^{152}Eu (open circles) are compared with the in-situ measurement method using the Ge detector (closed squares). Both the specific activities were in good agreement (except for ^{54}Mn). Thus, the net contact dose rate is available for estimating the specific activity of every radionuclide.

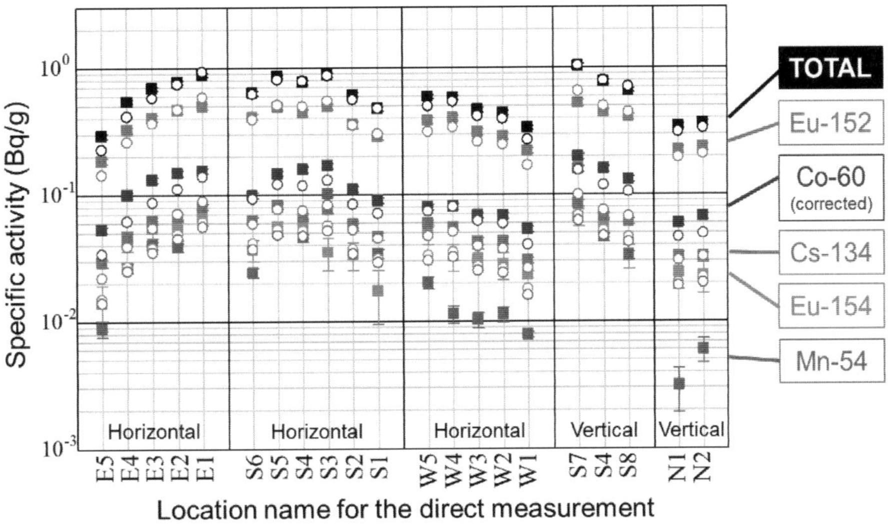

Figure 9. Comparison of the specific activities of ^{152}Eu, ^{60}Co (corrected), ^{134}Cs, ^{154}Eu, ^{54}Mn, and their sums in the concrete walls between by the in-situ measurement method using the Ge detector (closed squares) and by estimation from the net contact dose rates measured using the NaI survey meter (open circles).

4. Conclusions

The dominant radionuclides after ~2 years from the end of operation were found to be ^{152}Eu and ^{60}Co for the decommissioning of the concrete walls of a PET cyclotron room at the Medical and Pharmacological Research Center Foundation in Hakui, Ishikawa, Japan. Since the depth profile of specific activity was uniform approximately from the surface to a depth of ~20 cm, the specific activity could be determined by the in-situ determination method using a portable Ge detector.

There is a clear correlation between the total specific activity of γ-ray-emitting radionuclides and net contact dose rate. Therefore, the specific activity can be determined approximately by contact dose-rate measurement using a NaI survey meter. The specific activity ratios of ^{60}Co, ^{134}Cs, and ^{154}Eu for ^{152}Eu were independent of the location. The specific activity of each γ-ray-emitting radionuclide can also be determined approximately by contact dose-rate measurement using a NaI survey meter.

The in-situ measurement method is a powerful tool for the decommissioning of a PET cyclotron room because a contour map of the specific activity can be obtained in detail without destroying the structure.

References

[1] Masumoto K, Matsumura H, Bessho K and Toyoda A 2008 *J. Radioanal. Nucl. Chem.* **278** 449

[2] Wang Q B, Masumoto K, Bessho K, Matsumura H, Miura T and Shibata T 2004 *J. Radioanal. Nucl. Chem.* **262** 587

[3] Kusumoto K, Mizuno K and Iwata R 1992 *Proceedings of the IVth International Workshop on Targetry and Target Chemisty (TARGETRY'91), September 9-12, 1991, PSI Villigen, Switzerland, in PSI-Proc.* **92-01**

[4] Firestone R B 1996 *Table of Isotopes,* 8th ed., John Wiley and Sons, Inc., New York

[5] Venkataraman R, Bronson F, Atrashkevich V, Field M and Young B M 2005 *J. Radioanal. Nucl. Chem.* **264** 213

[6] Bronson F L 2003 *J. Radioanal. Nucl. Chem.* **255** 137

IOP Conf. Series: Journal of Physics: Conf. Series **1046** (2018) 012017 doi:10.1088/1742-6596/1046/1/012017

Evaluation of induced activity in various components of a PET-cyclotron

A. Toyoda[1], G. Yoshida[1], H. Matsumura[1], K. Masumoto[1], T. Nakabayashi[2], T. Yagishita[2] and H. Sasaki[2]

[1]Radiation Science Centre, High Energy Accelerator Research Organization (KEK), Oho Tsukuba, Ibaraki, Japan

[2]Japan Environmental Research Co., Ltd., 1-24-6 Nishi-Shinjuku, Shinjuku, Tokyo, Japan

akihiro.toyoda@kek.jp

Abstract. For decommissioning a cyclotron facility, it is important to evaluate the induced activity of the various components of the cyclotron; however, activation of the metal components has been rarely investigated. In this study, two types of cyclotrons were examined; one is a proton acceleration type using a deflector, and another is a hydride ion (H^-) acceleration type using a carbon stripper foil for beam extraction to the target port. The samples were obtained from various metal components such as the yoke, sector magnet, coil, and vacuum chamber by the core boring method, and the depth distribution of the radioactivity was determined via a germanium semiconductor detector. The activities of ^{54}Mn and ^{60}Co were detected from the surface to a deeper site of the yoke and sector magnet. Most of the observed activities of the cyclotron components were higher than the clearance levels, suggesting that a clearance system should not be applied to the yoke and sector magnet. In the case of a high-activity sample, we have to wait for 30 years to reach the clearance level.

1. Introduction

In Japan, various small-scale cyclotron facilities had been established for the generation of radiopharmaceuticals for use in positron emission tomography (PET). However, some facilities are aging, and it is expected that the number of systems to be decommissioned will drastically increase in the future. Though radioactive waste generation during the decommissioning of a cyclotron must be strictly controlled, extremely low-level radioactive wastes cannot be regarded as radioactive. Thus, such wastes can be treated as ordinal materials in a clearance system in Japan [1], and borderlines are configured for each nuclide (clearance level) [2]. For progressing the decommissioning of a cyclotron efficiently, a clearance system should be extensively utilized and the amount of radioactive waste generation should be reduced. Therefore, the methods to evaluate the activation of the components and facilities and to analyse samples must be clarified for the various types of cyclotron facilities.

In this regard, we examined the activation of two types of cyclotrons; one is a proton acceleration type using a deflector, and the other is a hydride ion (H^-) acceleration type using a carbon stripper foil for beam extraction to the target port. In both types, radionuclides get accumulated in the metal components during operation (i.e., radiopharmaceutical generation). It is already known that ^{60}Co and

^{54}Mn are generated in the metal components of a cyclotron, such as the yoke, sector magnet, vacuum chamber, and coil [3]; however, their radioactivity and distribution have not been comprehended yet. In this study, we examined the radionuclides that were induced in the metal components of a cyclotron and determined the depth and area distribution to obtain the fundamental data for conducting decommissioning.

2. Experimental procedures

2.1. Cyclotrons for sampling

The specifications of the two types of cyclotrons that were examined in this study are summarized in Table 1, and their diagrammatic outlines are shown in Figure 1. BC1710 that is manufactured by Japan Steel Works (JSW) is a proton acceleration type cyclotron and is not self-shielded. It was installed in a public hospital in the Kyoto prefecture, and was used from 2003 to 2016 for a total operation time of 3400 h. The principal parts of this cyclotron and its materials are as follows: yoke (iron), vacuum chamber (stainless steel), and coil (copper). We took samples from this cyclotron on 25 July 2016.
CYCLONE 10/5 that is manufactured by Ion Beam Applications (IBA) Belgium is a H^- acceleration type cyclotron with self-shielding. This type accelerates hydride ions and strips all the electrons using a carbon foil just before injection into a target. It was installed in a university hospital in the Mie prefecture, and was used from 2003 to 2011 for a total operation time of approximately 2600 h. The principal parts of this cyclotron and their materials are as follows: yoke (iron), sector magnet (iron), coil (copper), and vacuum chamber (aluminium). We took samples from this cyclotron on 24 May 2016. Both models were mainly used to produce ^{18}F as a radiopharmaceutical for PET.

Table 1. Specifications of the cyclotrons

	BC1710	CYCLONE 10/5
Manufacturer (Country)	JSW (Japan)	IBA (Belgium)
Self-shield material	None	Concrete
Maximum energy	17 MeV (p), 10 MeV (d)	10 MeV (H^-), 5 MeV (D^-)
Maximum current	50 μA (p), 50 μA (d)	60 μA (H^-), 35 μA (D^-)
Operation time	4/2003–4/2016	3/2003–11/2011
Total operation time	3400 [h]	2598 [h]
Sampling date	25/7/2016	24/5/2016

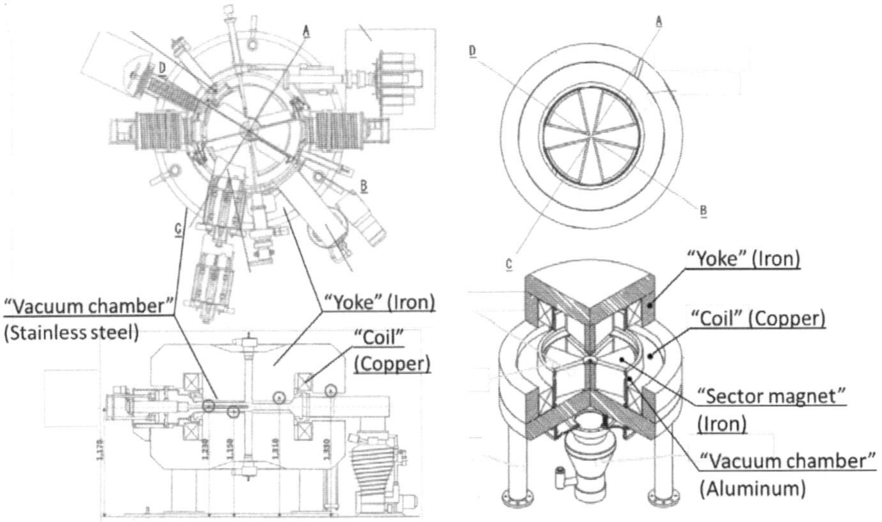

Figure 1. Diagrammatic outlines of BC1710 (left) and CYCLONE 10/5 (right)

2.2. Core boring and sample preparation

Sample preparation processes are shown in Figure 2. Both the cyclotrons were divided into individual components, and core samples were extracted from the yoke, sector magnet, coil, and vacuum chamber by a drill (core boring method [4,5]). First, to easily assess the activation level of each part, the surface dose rate of the core samples was measured via a sodium iodide (NaI) scintillation survey meter (TCS-161, Hitachi). The measurements were performed along the axial direction in steps of 5 cm from the surface. We adopted the median of each unit as the measuring point (i.e. 2.5 cm, 7.5 cm, 12.5 cm…). The time constant of the survey meter was 10 s. After the dose rate measurement, the samples were sliced into 1-cm test pieces. Each test piece was packed in a plastic container (U8 type) and characterized by γ-ray spectrometry. Such core sampling from the metal components of a small-scale medical purpose cyclotron and similar activity measurements have been rarely performed.

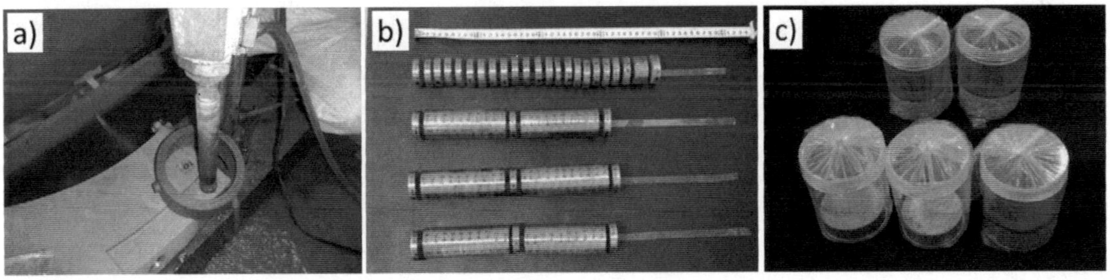

Figure 2. Principal processes of sample preparation, a) core samples (diameter: 5 cm, length 3–36 cm) were extracted from the metal components of the cyclotron using a core drill, b) extracted samples were sliced into 1-cm-thick test pieces by a tip saw after the dose rate measurement via a sodium iodide (NaI) scintillation survey meter, and c) each test piece was packed in a plastic container (U8 type) for characterization by γ-ray spectrometry.

2.3. γ-ray spectrometry

We performed γ-ray spectrometry, using a germanium semiconductor detector (GMX-20195-S, ORTEC), for the test pieces prepared by the method described in the previous sub-section. The detector head was shielded by an inner sleeve (5 mm of copper and 5 mm of acrylic) and an outer sleeve (11 cm of lead and 5 cm of iron). The measurement time was 10000 s for each sample. The detection efficiency was determined using the standard radioisotopes (RI) solution of the Japan

Radioisotope Association (JRIA). Simultaneously, the self-absorption by the sample and sum-peaks were corrected. The activity of each nuclide was corrected on the sampling date.

3. Results and discussion

3.1. Dose rate of a core sample at each sampling point

The dose rates of the core samples are summarized in Tables 2 and 3. The background of the measurement is 0.06 μSv/h, and the detection limit is 0.09 μSv/h. The values presented in the tables include the background. From the results of BC1710, we find that the dose rates of the yoke (Nos.1–10 in Table 2) are much higher than the background level at all the sampling points until a depth of 20 cm. In contrast, for CYCLONE (Table 3), the dose rates are almost similar to the background level at numerous sampling points. It was expected for the short-lived nuclides to have decayed because 4.5 years had elapsed since the end of operation. The dose rates of the vacuum chamber of BC1710 (Nos.11–13 in Table 2) were quite high because of the presence of short-lived nuclides such as ^{51}Cr, ^{57}Co, and ^{58}Co that were generated by nuclear reactions with Cr and Ni. These three samples were only 3.3 cm in length, and thus, a median length of 1.65 cm was adopted as the measurement point.

Table 2. Dose rates of the core samples extracted from BC1710 (μSv/h)

Sample name	Depth from the surface					
	2.5 cm	7.5 cm	12.5 cm	17.5 cm	22.5 cm	27.5 cm
Yoke (Nos.1–10)						
No.1	0.21	0.19	0.14	0.11	0.09	0.08
No.2	0.31	0.35	0.29	0.20	0.14	0.11
No.3	0.19	0.17	0.14	0.12	0.11	0.10
No.4	0.33	0.28	0.21	0.14	0.12	0.10
No.5	0.22	0.19	0.14	0.11	0.10	0.08
No.6	0.23	0.22	0.16	0.12	0.10	0.09
No.7	0.40	0.35	0.23	0.15	0.10	
No.8	0.23	0.23	0.18	0.15	0.12	0.09
No.9	0.36	0.31	0.23	0.16	0.11	0.10
No.10	0.21	0.20	0.15	0.12	0.10	0.08
	Depth from the surface					
	1.65 cm					
Vacuum chamber (Nos.11–13)						
No.11	1.40					
No.12	2.40					
No.13	1.05					

Table 3. Dose rates of the core samples extracted from CYCLONE 10/5 (μSv/h)

IOP Conf. Series: Journal of Physics: Conf. Series **1046** (2018) 012017 doi:10.1088/1742-6596/1046/1/012017

Sample name	Depth from the surface					
	2.5 cm	7.5 cm	12.5 cm	17.5 cm	22.5 cm	27.5 cm
Sector magnet (Nos. 1-8):						
No.1	0.21	0.22	0.20	0.19	0.17	0.13
No.2	0.06	0.06	0.06	0.06	0.06	0.06
No.3	0.09	0.10	0.09	0.08	0.08	0.07
No.4	0.09	0.09	0.09	0.09	0.08	0.07
No.5	0.17	0.18	0.17	0.16	0.14	0.12
No.6	0.06	0.06	0.06	0.06	0.06	0.06
No.7	0.09	0.10	0.10	0.10	0.09	0.08
No.8	0.08	0.09	0.09	0.09	0.09	0.07
Yoke (Nos. 9-24):						
No.9	0.06	0.06	0.06	0.06		
No.10	0.06	0.06	0.06	0.06		
No.11	0.11	0.10	0.08	0.07		
No.12	0.06	0.06	0.06	0.06		
No.13	0.06	0.06	0.06	0.06		
No.14	0.11	0.10	0.08	0.06		
No.15	0.06	0.06	0.06	0.06		
No.16	0.06	0.06	0.06	0.06		
No.17	0.06	0.06	0.06			
No.18	0.22	0.18	0.11			
No.19	0.07	0.06	0.06			
No.20	0.06	0.06	0.06			
No.21	0.06	0.06	0.06			
No.22	0.06	0.06	0.06			
No.23	0.25	0.21	0.13			
No.24	0.06	0.06	0.06			
Coil (Nos.25–32):						
No.25	0.08	0.06	0.06	0.06		
No.26	0.06	0.06	0.06	0.06		
No.27	0.06					
No.28	0.06					
No.29	0.08					
No.30	0.06					
No.31	0.06					
No.32	0.06					

3.2. Depth distribution of the nuclides induced in the cyclotron metal components

IOP Conf. Series: Journal of Physics: Conf. Series **1046** (2018) 012017 doi:10.1088/1742-6596/1046/1/012017

From the results of the dose rate measurements recorded by the survey meter, it was revealed that radioactivity was induced in the deep sites of the components. In this section, we discuss the γ-ray spectrometry that was performed for the test pieces (sliced core samples) to investigate the type of nuclides generated in the cyclotron. The lengths of the yoke and sector magnet being suitable to determine the depth distribution of the nuclides and the corresponding samples were chosen for the spectrometry measurements. The sampling points for the yoke of BC1710 and depth distribution of the nuclides at each point are shown in Figure 3. The bottom side of the yoke is depicted as the area shaded in blue in Figure 3. The sample cores were extracted along the red lines directed from the centre to the outside. The horizontal axis denotes the core sample depth, from the centre to the outside direction. The vertical axis corresponds to the specific activity of a particular nuclide at each depth. As shown in Figure 3, ^{54}Mn (green diamond) and ^{60}Co (red triangle) are detected at all the sampling points. Although in a small amount, ^{59}Fe (yellow circle) is also detected. The specific activity of ^{54}Mn rapidly decreases with the depth, whereas that of ^{60}Co decreases relatively slowly. ^{54}Mn was generated by the fast neutron, whereas ^{60}Co was mainly generated by the thermal neutrons. Such differences in the production process could influence the activity at each depth. At sampling points Nos. 6–8 and 10, the specific activity of ^{60}Co decreases with the depth, though there is a slight increase at the end point of the core sample. This is caused by the reflected neutrons generated by the shield wall which is just outside the yoke.

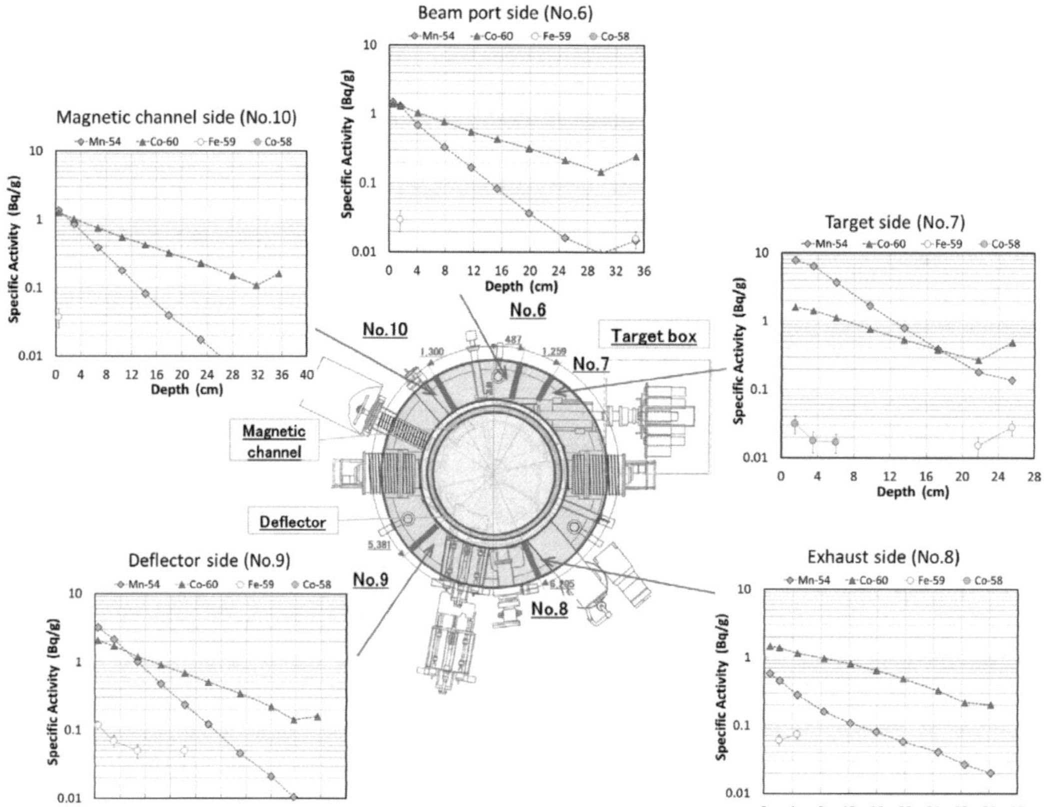

Figure 3. Depth distribution of the specific activity of the principal nuclides at the sampling points (BC1710, bottom side of the yoke). The yoke is the area shaded in blue. The sample cores were extracted along the bold red lines directed from the centre to the outside.

The sampling points for the yoke of CYCLONE 10/5 and depth distribution of the nuclides at each point are shown in Figure 4. The second stage of the yoke is depicted as the area shaded in blue in Figure 4, whereas the red lines depict the sample cores. In all the sampling points shown in Figure 4, ^{60}Co is detected, whereas ^{54}Mn is mainly detected in the target side (No.23). Similar to BC1710, the specific activity of ^{54}Mn rapidly decreases with depth, whereas that of ^{60}Co decreases more slowly. The activity levels of ^{54}Mn and ^{60}Co are similar to those of the yoke of BC1710 at the point close to the target (No.23 of CYCLONE and No.7 of BC1710). In comparison, in the other areas, the activity levels are much lower than those of BC1710, even when considering the attenuation of the activity by a time-lapse. This result reflects that the beam loss in the H$^-$ acceleration type cyclotron is much smaller than in the proton acceleration type.

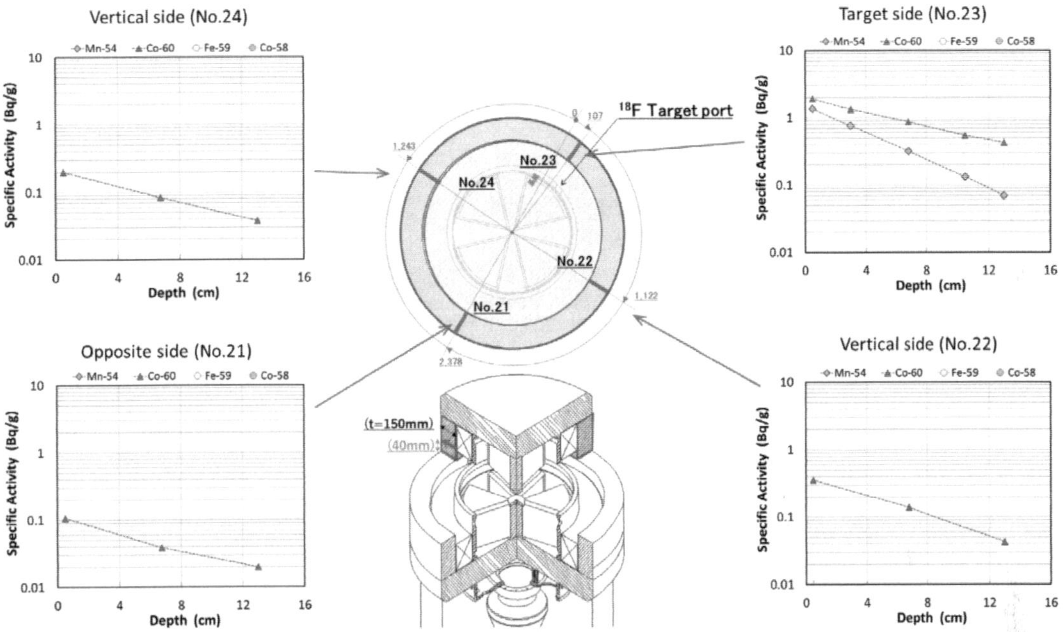

Figure 4. Depth distribution of the specific activity of the principal nuclides at the sampling points (CYCLONE 10/5, Second stage of Yoke). The yoke is the area shaded in blue. The sample cores were extracted along the bold red line from the centre to the outside direction.

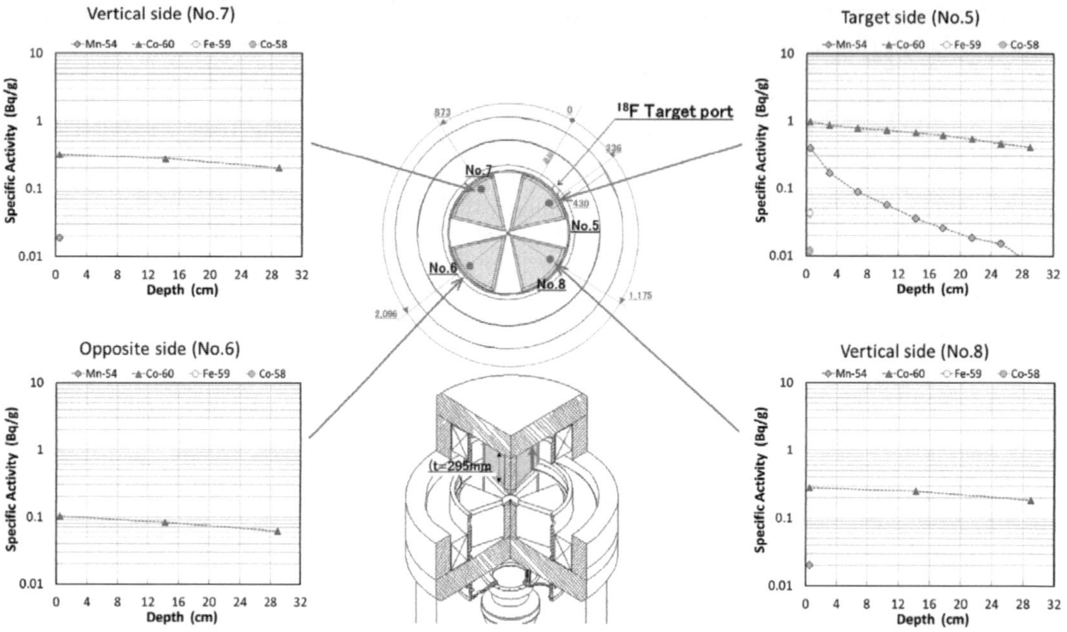

Figure 5. Depth distribution of the specific activity of the principal nuclides at the sampling points (CYCLONE 10/5, top side of the sector magnet). The sector magnet is the area shaded in green. The sample cores were extracted from the direction of the bottom to the top at each red dot.

The sampling points of the sector magnet in CYCLONE 10/5 and depth distribution of the nuclides at each point are shown in Figure 5. The sector magnet is the area shaded area in Figure 5 Same as the results for the yoke, ^{60}Co is detected at all the sampling points in Figure 5, and ^{54}Mn is mainly detected on the target side (No.5). The specific activity of ^{54}Mn rapidly decreases with the depth. This is the same trend as for the yoke. As a unique trend of the sector magnet, the specific activity of ^{60}Co only slightly decreases with the sample depth. It is implied that the thermal neutrons were distributed in the entire body of the sector magnet during the operation. The activity level of ^{60}Co is almost same to that of the yoke in CYCLONE 10/5, whereas the level of ^{54}Mn is lower.

3.3. Verification of the activity of the cyclotron components by the clearance level

The ratio of the observed activity to the clearance level (D/C) was considered for the results presented in section 3.2. The clearance level of both ^{54}Mn and ^{60}Co is 0.1 Bq/g, and the specific activity of the components must be below the clearance level to treat the radioactive wastes as normal wastes (i.e., D/C < 1). In the case of BC1710 (Figure 3), D/C is larger than 1 at all the sampling points and depths. Furthermore, in case of CYCLONE 10/5 (Figures 4 and 5), D/C is larger than 1 at various sampling points. The D/C ratio is marginally lower than 1 at only two sampling points (Nos. 6 and 21). In the case of the highest activity sample, it is expected that 30 years or more will be necessary to reduce the activity level below the clearance level. From these facts, it is found that the cyclotron main components (yoke and sector magnet) cannot be treated as normal wastes, regardless of the acceleration particles. These must be controlled as radioactive wastes.

4. Conclusions

We analysed the core samples from two types of small-scale medical purpose cyclotrons. We examined the radioactivity of the nuclides generated in the metal components of the cyclotrons and determined the depth distribution. In all the samples, ^{54}Mn and ^{60}Co were detected as the principal nuclides. ^{54}Mn, which was generated by the fast neutron reactions, was mainly detected close to the target port and rapidly decreased with the depth. In contrast, ^{60}Co, which was generated by the thermal neutrons in all the parts, slowly decreased with depth. The difference, which was attributed to the acceleration particles, was very small. The metal components of the cyclotron such as the yoke and sector magnet should be treated as radioactive wastes because their activity is expected to be much higher than the clearance level.

References

[1] https://www.nsr.go.jp/activity/regulation/nuclearfuel/haiki4.html (Japanese only)
[2] Okoshi M 2002 *Jpn. J. Health Phy.* **37**(3) pp197–207
[3] Ito K, Nakata Y, Matsuda H, and Sato N 2011 *Kaku Igaku.* **48**(2) pp109–119
[4] Saito K, Takahashi Y, Yamaguchi I, Kimura K, Kanzaki T, Shimada H, Otake H, Oriuchi N, and Endo K 2009 *Jpn. J. Med. Phys.* **29**(2) pp29–34
[5] Masumoto K, Ohtsuki T, Kasagi J, Izumi Y, and Yamato I 1999 *Jpn. J. Health Phy.* **34**(2) pp151–160

IOP Conf. Series: Journal of Physics: Conf. Series **1046** (2018) 012018 doi:10.1088/1742-6596/1046/1/012018

Assessment of environmental consequences of the normal operations of the ESS facility

D Ene[1], R Avila[2], T Hjerpe[2], D Bugay[2], K Stenberg[2]

[1] ESS ERIC Tunavägen 24, Lund, 221 00 Lund, Sweden
[2] FACILIA AB Gustavslundsvägen 151C, 167 51 Stockholm, Sweden

Daniela.ene@esss.se

Abstract. As other accelerator based facilities, the European Spallation Source ESS facility will interact with the environment. The Swedish legislation requires a demonstration that the sum of the doses resulting from the exposure of any member of the public to ionizing radiation dose does not exceed the specified limit of 50 μSv/year. A radiological assessment has been produced to provide that demonstration. This evaluation was based upon the actual status of the ESS design. A graded approach was adopted through over the assessment allowing estimating dose for all radionuclides and exposure pathways, but the degree of detail in the assessment depend upon their relative radiological importance. The total dose was obtained making the sum of the contribution of all-important radionuclides treated realistically with that of all screened out radionuclides, derived by means a conservative method.

1. Assessment approach

For the derivation of the release-to-dose-factors and estimation of radionuclide concentrations in the environment, ESS is using PREDO (PREdiction of DOses from normal releases of radionuclides to the environment) methodology.

A graded approach was adopted through over the assessment allowing estimating dose for all radionuclides and exposure pathways, but the degree of detail in the assessment depend upon their relative radiological importance. Firstly, the relative importance of the dose contributions was assessed by comparing doses calculated based on IAEA SRS19 [1] screening approach against predefined dose value of 0.1 μSv/year. In addition, it was required that the sum of the doses of screened out radionuclides is also below the exemption level of 10 μSv/year. In the second step, for all radionuclides previously identified having significant dose contribution, realistic dose factors were derived using dispersion and radio-ecological realistic models.

Finally, the total dose was obtained making the sum of the contribution of all-important radionuclides treated realistically with that of all screened out radionuclides, derived by means the conservative method. The assessment of radiological impacts from releases of radioactive substances to the environment was carried out in two steps: i) estimation of the source term consisting of calculation of activities of radionuclides that can be released annually from the ESS facility and ii) applying dispersion and transport models to calculate radionuclide concentrations in the environment and dose models to calculate doses to reference groups from the population around the site.

The evaluation of exposures to impacted environmental media considers three main pathways: i) airborne releases of radionuclides through ventilation outlets ii) liquid discharges of radionuclides to the sewage system and downstream surface water (rivers and sea), iii) migration of radionuclides with groundwater following activation of the surrounding soil.

2. Releases of radioactive substances to the atmosphere

The releases from the main release point, the main stack, are presents in this work. This is the point of continuous releases from the accelerator tunnel and the target station as well as of periodical short-term releases from the hot cell, occurring during the dismantling and processing of the target system and other obsolete components.

2.1. Source term evaluation

The estimated source term is presented in the following sections, according to the origin of releases.

2.1.1. Tunnel of accelerator. The radionuclide concentrations in the tunnel have been calculated [2] from the radionuclide production yields considering the time evolution of the air inside the tunnel for one air exchange rate [h^{-1}] the baseline parameter of the ventilation circuit. The results of the calculations integrating over 6000 h of continuous operations are provided in figure 1.

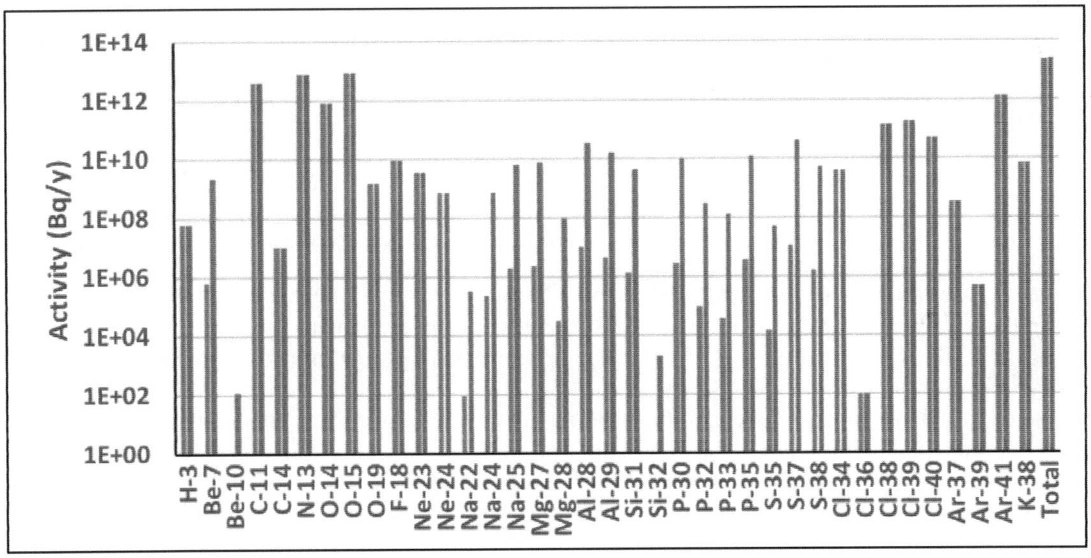

Figure 1. Source term for airborne release during operation. In blue, the effectiveness of the hepa filter (99.97%) is applied for aerosols.

As can be seen in the figure, the source term to the stack takes into consideration the filter effect (blue columns). It was concluded from this analysis that during operation, the ventilation system will extract about 22.6 TBq of air radioactivity per year given mainly by the short lived gases as ^{13}N, ^{11}C, $^{14,15}O$ and ^{41}Ar. Therefore the filter effect upon the total released activity is small (less then 2%).

2.1.2. Target station. A detailed analysis of the potential releases to the environment from the main target systems was performed in [3]. Within the Target Station there are four important contributors to radioactive releases in locations where the ventilation system will be implemented: i) helium cooling loop (HeL) of the tungsten (W) target, ii) hot cells (HC) facility , iii) gas-liquid separator (GLS) tanks of the main cooling water circuits and iv) helium within the shielding monolith. In this work only first two components of the source term are addressed.

IOP Conf. Series: Journal of Physics: Conf. Series **1046** (2018) 012018 doi:10.1088/1742-6596/1046/1/012018

2.1.2.1 Helium cooling loop (HeL) is the most critical zone where spallation products created within the W target have the potential to be released. Therefore a purification system based on getters was necessary to be inserted into the helium loop. The HeL contains also special filters for W dust that is formed by ablation, see figure 2. The volatiles and dust that will leak from the purified helium through the pipes will be extracted by the ventilation system from the rooms housing the loop. According to the current knowledge, there are three potential mechanisms of radionuclide releases from the target within HeL: i) direct ejection or sputtering due to spallation reactions, i) diffusion and iii) ablation. The species assumed to be released from the W target within the helium loop are almost all spallation products of the W target. The block diagram in figure 3 shows the possible locations where activity can reside, and the ways in which it can move. "Surfaces" is a general term describing any place within the HeL where dust may be plated.

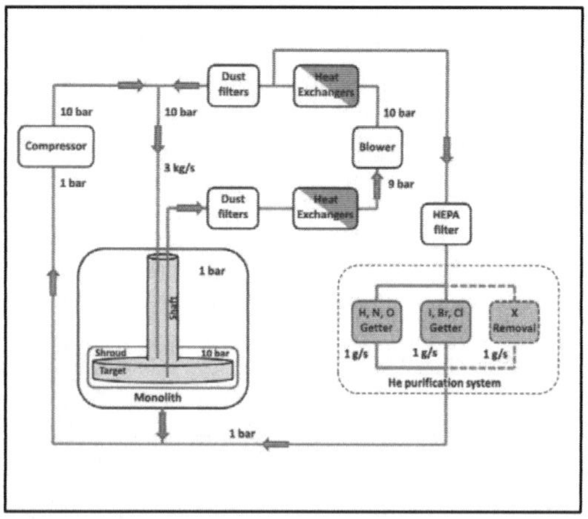

Figure 2. Schematic view of the helium cooling loop of the tungsten target.

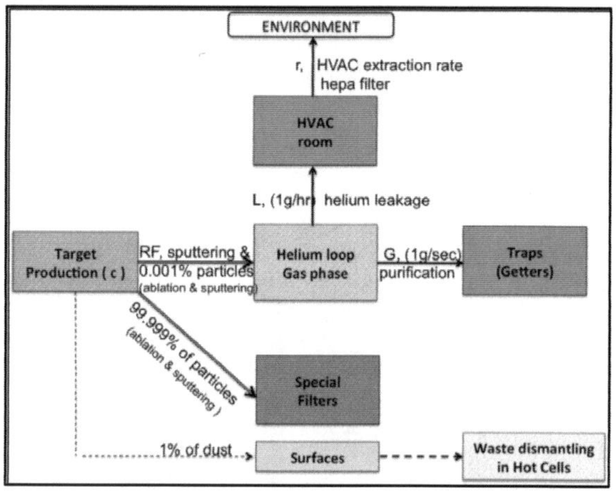

Figure 3. Compartment model for developing the analytic expressions governing radioactivity releases from the helium cooling loop.

Analytical solutions to the differential equations developed based on this compartment model governing radionuclide turnover in the system as well as the parameters values used in the calculations are provided in [3]. The obtained source term amounts at 6.32E+11 Bq. It contains more than 1050 radionuclides, which were grouped according to their properties: gases, volatiles, semi-volatiles and particulates or metalloids. It was found, for example, that the annual release of the iodine isotopes is 1.4E+09 Bq, while of the noble gases is 6.6E+09 Bq.

2.1.2.2 *Hot cells (HC) facility* is the potential origin of the following releases to the environment arising during the processing:

1. Tungsten (W) dust (pasted-up on the surfaces of the structures and pipes) that can be spread and arise during the cutting process;
2. Stainless steel dust arising from machining of the obsolete components (target shaft and shroud, moderator reflector twister, pipes, plugs, others).

The ventilation system associated to HC has at least two levels of hepa filtration: i) local and ii) at the stack. In the present estimates credit is taken for at least one level of filtration. It was assumed that all W dust accumulated on the surfaces (see figure 3) during five years of irradiation will be spread within the processing HC and released via ventilation system to the stack. Firstly, the accumulated activity at

the end of the 5th annual run was estimated and thereafter the annual release to the environment, considering a continuous uniform dismantling over two months, was derived.
The annual release of the W dust is about 3.1E+7 Bq for the worse case, when only the first level of the ventilation filtration located at the HC level is accounted for. Annual release is about 9.2E+3 Bq when considering also the effectiveness of the second level of hepa filtration at the stack. The annual release is completed by adding tritium assumed to be released from both W target bricks and W dust during dismantling, about 2.6E+11 Bq. It is further assumed that other gases and volatiles in the W dust stay attached and behave as dust.
The radioactive inventory of the stainless steel dust (SS dust) produced in the HC during cutting the shaft and the shroud of the target system was calculated in two steps. Firstly, the radioactive inventory of each of the items, the shroud and the shaft, to be cut within the HC was calculated. Secondly, the amount of SS dust that might arise during the cutting process was estimated. And finally, the annual release was derived as above using the maximum of SS dust that can be produced during processing.
The estimated release of the SS dust is 4.6E+08 Bq in case if only one filtration level is accounted and 1.4E+05 Bq if subsequent filtration at the stack is also accounted for. The source term accounts for 2.0E+08 Bq of tritium, assumed to be released from the SS dust. The remaining tritium in the SS dust is assumed to stay attached and behave as dust.

2.2. Doses from releases to the atmosphere

2.2.1. Screening study for releases from the main stack. From calculations with the IAEA screening models [1] conservative release-to-dose-factors were obtained for all radionuclides that can be potentially released from the main stack. These release-to-dose-factors were then multiplied by the total releases from the main stack, to obtain a conservative estimation of the annual dose from each released radionuclide. Both continuous and short-term releases were included in the in the release rate used for the screening. It was found that only for 16 radionuclides the screening doses were above the screening criteria of 0.1 μSv/y. Consequently they were further treated by means realistic models. The total screening dose from these 16 radionuclides is about 25 μSv/y. The total screening dose rate of the remaining radionuclides that were screened out is 0.54 μSv/y.

2.2.2. Dose estimates for continuous releases from the main stack. The following dose exposure pathways were considered in the realistic models for atmospheric releases: i) immersion in the plume; ii) exposure to deposited radionuclides; iii) ingestion of food from a garden plot; iv) ingestion of crops from a cropland area; v) ingestion of meat and milk; vi) ingestion of forest food; vii) ingestion of water and viii) ingestion of freshwater food. The realistic doses were calculated for representative land use objects: residential areas, croplands, pasturelands, forests, Kävlinge river subcatchment area and discharges to surface waters: the Källby sewage treatment plant and the part of Höje river which flows downstream Källby plant. Three age groups were considered: Infant, Child and Adult. In the following, results are presented for averaged Adult and the other groups are treated in the uncertainty analysis. Two cases of release rates were considered for the estimation of the annual doses:

1. Pessimistic estimate of the release rate. This is the most pessimistic case, assuming the maximum that the facility can potentially release to the environment. It means no credit is given to the filtration equipment (all hepa filters are down) at the main stack. Only the clean-up and abatement equipment located inside the buildings are considered in function. This source term would never be approached during normal operation.
2. Realistic estimate of the release rate. In this case all abatement equipment works optimally, including the filtration system.

Annual Dose (Sv/y) to the Average Adult and contributions to the annual dose from different radionuclides for the two analysed release cases are given in Table 1.

Table 1. Annual effective dose to the reference Averaged Man. Results from linac, Target Station-helium cooling loop and hot cells.

	Component	Released Activity (Bq/y)		Effective dose (μSv/y) Screening approach		Effective dose (μSv/y) Realistic approach	
		Pessimistic release	Optimal release	Pessimistic release	Optimal release	Pessimistic release	Optimal release
	Linac	2.26E+13	2.25E+13	1.16E+01	1.02E+01	5.44E-01	2.82E-01
Continuous release	HeL:						
Continuous release	Iodine	1.35E+09	1.35E+09	1.20E+01	1.20E+01	5.57E-02	5.57E-02
Continuous release	Other halogens	5.94E+07	5.94E+07	1.98E-03	1.98E-03	1.98E-03	1.98E-03
Continuous release	H-3	6.24E+11	6.24E+11	1.08E+00	1.08E+00	5.60E-03	5.60E-03
Continuous release	Gases (O, N, CO)	7.33E+03	7.33E+03	7.53E-09	7.35E-09	3.83E-09	3.83E-09
Continuous release	Noble gases	6.60E+09	6.60E+09	1.28E-03	1.28E-03	1.28E-03	1.28E-03
Continuous release	Metaloids_1	2.72E+05	8.17E+01	9.20E-04	2.76E-07	3.26E-07	2.45E-07
Continuous release	Metaloids_2	5.29E+07	1.59E+04	2.06E-01	6.19E-05	1.02E-04	6.20E-05
Continuous release	Total target	6.32E+11	6.32E+11	1.33E+01	1.31E+01	6.47E-02	6.47E-02
Continuous release	**Total at stack**	**2.33E+13**	**2.31E+13**	**2.49E+01**	**2.33E+01**	**6.79E-01**	**3.47E-01**
Short term release	Dismantling in the Hot Cells						
Short term release	W dust	6.24E+11	6.24E+11	1.26E+00	1.08E+00	3.47E-02	2.92E-02
Short term release	SS dust	6.60E+08	2.01E+08	7.44E-01	5.71E-04	1.37E-01	5.05E-05
Short term release	Total	6.25E+11	6.24E+11	2.00E+00	1.08E+00	1.72E-01	2.93E-02
	TOTAL each 5 years	**2.39E+13**	**2.37E+13**	**2.69E+01**	**2.44E+01**	**8.51E-01**	**3.76E-01**

The total annual effective dose obtained for both variants are below 1 μSv/year and is dominated by N-13, C-11, Ar-41, O-15, I-125 and H-3, see figure 4.

The reduction effect of the filtration system on the total doses from continuous releases is marginal, since the total doses are dominated by radionuclides that are released in gas form and are therefore not retained in the filters.

2.2.3. Dose estimates for short term releases from the main stack. For short term releases, the release-to-dose-factors were calculated using a repeating pulse as release rate. The pulse was defined as a constant release of length 2 /12 y and amplitude 12/2 Bq/y occurring repeatedly every 5th year. In this way, the release rate corresponds to 1 Bq of released activity during the release period. release-to-dose-factors were calculated the following two-month-long release periods: January and February, March and April, May and June, July and August, September and October, November and December. Interception of deposited radionuclides onto plant surfaces was assumed for all release periods except those occurring in the winter (January and February, November and December).

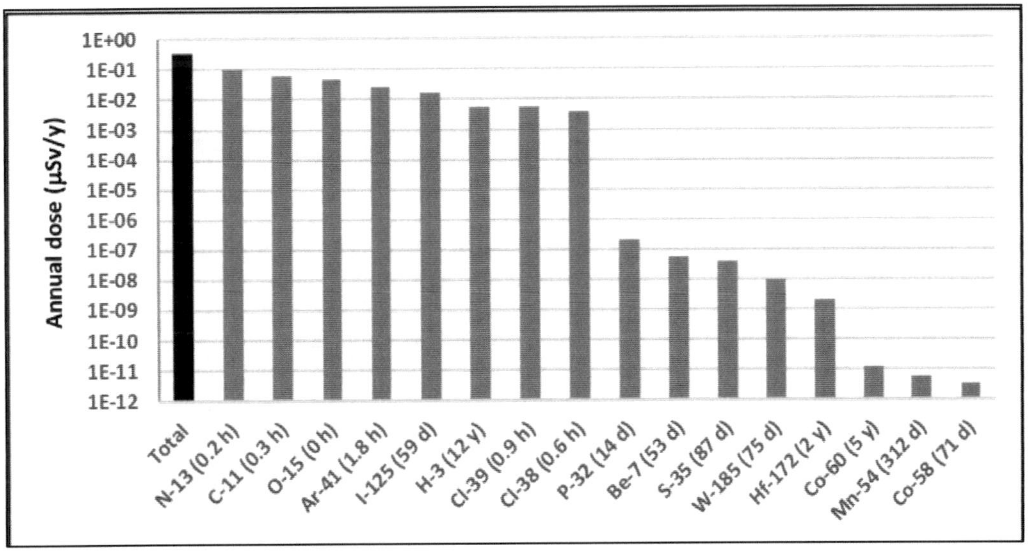

Figure 4. Main stack continuous release. Total annual effective dose and contribution of the dominant 16 radionuclides (80%).

Table 1 contains the maximum annual doses corresponding to the period September-October and contributions from the two analysed release cases. The total dose from short term releases from the main stack is small in comparison to the contribution from continuous releases. The reduction effect of the filtration system on the total doses from short term releases from the main stack is negligible, since the total doses are dominated by releases of H-3 occurring during dismantling of the obsolete target. While for the dust, which is retained in the filters the effect is about three orders of magnitude.

The results of realistic dose calculations for airborne releases of radionuclides from the main stack show that annual doses to members of the public are well below the Swedish regulatory constraint of 100 µSv/year and even below the exemption level of 10 µSv/year.

3. Discharge to the sewage system

All radioactive wastewater produced during the operation (run cycles and maintenance) will be sent to the waste facility (via pipes or tanks) for treatment. If the activity levels of the radionuclides in the treated wastewater meet the requirements, then the waste water can be discharged to the sewage plant in Lund. Radionuclides potentially contained in the wastewater originate from: i) activation of the water, and ii) contamination of water with corrosion products and dust. The breakdown of radionuclides with potential to be released to the sewer system from the waste facility is given in [5].
Radionuclides from the ESS facility discharged to surfaces waters are assumed to be directly received by the Källby sewage treatment plant. After treatment, the water is released to the downstream freshwater receptor Höje river. Further downstream, the marine basin (Lomma bay) is considered the final receptor of the discharges.
The assessment accounts for exposure to 1) workers in the sewage plant that are exposed to radionuclides in the non-treated water or sewage sludge and 2) the general public exposed, either directly (through consumption) or indirectly (though use of water for irrigation or for feeding cattle).
Workers in the Källby sewage treatment plant are assumed to be exposed by: i) external irradiation from radionuclides in the water or sludge and ii) from internal exposure though inhalation of re-suspended radionuclides and inadvertent ingestion of sludge.
The general public is assumed to be exposed to radionuclides by the following exposure pathways: i) ingestion of freshwater and freshwater food from the Höje river, ii) ingestion of marine food from the

Lomma bay, iii) Marine activities in the Lomma bay (boating, swimming and sunbathing), iv) exposure to radionuclides accumulated on the ground as result of irrigation with water from the Höje river, v) ingestion of irrigated garden foods, vi) ingestion of meat and milk from cattle consuming contaminated water. Because quantification of the wastewater discharge rates from the waste facility outlet was not available the dose calculations could not be carried out. Therefore, for all identified radionuclides realistic release-to-dose-factors were derived using the models described in [6]. The realistic release-to-dose-factors were further used to derive discharged limits (reference discharge levels) for each radionuclide that could be potentially discharged to the sewage system. In deriving these limits, all potential exposures of sewage workers and the general public have been taken into account by dividing the assumed dose target 10μSv/y with the maximum of the Release-to-Dose-Factor for workers and general public. Reference [4] provides the obtained values of the reference discharge limits.

4. Migration of contaminants with the groundwater

4.1. Source term evaluation

The estimation of the source term of radionuclides generated by activation of the soil surrounding the accelerator tunnel is presented in [7]. This report derives detailed earth activation levels around the linac tunnel providing the results along the entire length of the machine. From all radionuclides derived in the soil only 11 were considered and justified as representative and used further as source term in the assessment: H-3, Be-7, Na-22, Na-24, P-32, S-35, Ca-45, Sc-46, Mn-54, Fe-55, Zn-65.

4.2. Groundwater modelling

The assessment of doses to reference persons from activation of soil and groundwater was carried out in two steps: i) groundwater flow modeling using MODFLOW code and ii) radionuclide transport modeling by means the NORMALYSA model library, see details about modelling in [8]. The groundwater flow model assumes steady state groundwater flow regime. Specific geological layers of the surrounding area and the associated hydrogeological and hydraulic parameters were used to create the 2D groundwater model of the flow from the tunnel to a well located at 300 m distance. The radionuclide transport is based on water fluxes and flow rates calculated in the step 1. The conceptual model of the radionuclide transport from linac to the well is shown in figure 5.

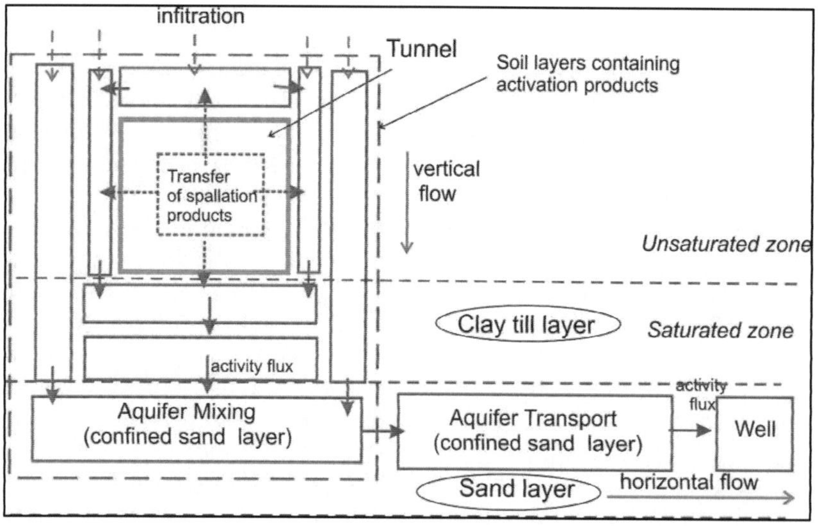

Figure 5. Conceptual model of radionuclide transport process to groundwater from the linac tunnel.

This model accounts for production of radionuclides in a sequence of soil layers surrounding the tunnel with a total thickness of 6 m. Radionuclide transport in the unsaturated and saturated soil zone was modeled using the advection-dispersion transport equation accounting for sorption described by kd model (linear reversible sorption). Transport calculations were carried out for the higher mobility radionuclides, namely: H-3, Na-22 and S-35. These higher risk radionuclides were selected from the list of 11 spallation products in the course of a previous screening radionuclide transport analyses.
Two scenarios of radionuclides leaching from the activated soil around the linac tunnel to the subsurface were analyzed:

1. Scenario 1 assumes an impermeable soil cover (or membrane), situated on top of the tunnel, which will fail after the end of operations of the accelerator (at the 40th year), leading to leaching of the accumulated nuclides in the activated soil to groundwater;
2. Scenario 2 assumes that there is no protective cover present, and a chronic release occurs through the operation life time of the facility (40 years).

Radionuclide concentrations in well water (Bq/m^3) for the reference case obtained for the two scenarios considered are presented in Table 2.

Table 2. Maximum radionuclide activity concentrations (Bq/m^3) in the water of the well for the two scenarios considered.

Radionuclide	Scenario #1		Scenario #2	
	Reference kd*	Conservative kd**	Reference kd*	Conservative kd**
^{3}H	3.57E+04	3.57E+04	1.67E+04	1.67E+04
^{22}Na	1.45E-32	2.91E-11	1.28E-31	2.01E-10
^{35}S	1.21E-88	1.10E-43	2.19E-87	1.75E-42

*kd provided in [8]
**kd (= Reference Kd/10)

The results for both scenarios show that doses will be formed practically 100% by tritium. Transport of other radionuclides to the well is delayed by sorption, and their predicted concentrations in well water are negligible. The results of modeling show that tritium concentrations in well water are approximately one order of magnitude below the admissible level of 100 Bq/L.

4.3. Dose estimates
Uses of the well water for drinking and irrigation of a garden plot were considered in the dose calculations. The models described in [6] were used in the dose estimations. Annual doses (Sv/y) from exposures to groundwater contaminated by activation are shown in Table 3

Table 3. Annual dose rate (μSv/y) from exposures to groundwater from linac tunnel area. Values are given for a representative Average Adult.

Radionuclide	Scenario #1		Scenario #2	
	Ingestion of water	Total	Ingestion of water	Total
^{3}H	2.42E-01	2.45E-01	1.130E-01	1.15E-01
^{22}Na	1.74E-35	3.22E-35	1.54E-40	1.54E-34
^{35}S	3.51E-92	6.50E-92	6.34E-92	1.30E-34
Total	2.42E-01	2.45E-01	1.13E-01	1.15E-01

Values are given for a representative adult with average habit. The estimated doses are the highest for the scenario of accumulation of radionuclides and below 1 μSv/year, considering the use of well water for drinking and irrigation of a garden plot.

5. Summary and conclusion

Maximum annual dose for continuous release to the main stack from linac tunnel and helium coolant loop is less than 1 μSv/y and the major contribution (80%) is given: N-13, C-11, Ar-41, O-15, H-3, I-125, Cl-38,39, P-32, Be-7, S-35, W-185, Hf-172, Co-60, Mn-54 and Co-58. Maximum annual dose for short term release to the main stack is less than 0.1 μSv/y and the major contribution is given by H-3 release during dismantling. The source term to the waste facility outlet is not known therefore derived activity limits where derived for all 72 radionuclide with potential of discharging into the sewage system. Maximum annual dose due to the migration of the contaminant with the groundwater is less than 1 μSv/y and is due to the drinking water contaminated with ^{3}H.
Actual results of realistic dose calculations for airborne releases of radionuclides from the main stack show that annual doses to the public are well below the regulatory constraint of 100 μSv/y and even below the exemption level of 10 μSv/y.
Obtained results shall be completed with remaining contributors.
This report is conceived according the knowledge that ESS staff has in this stage of the project. The current data are estimations subjected to evolution and update.
An environment monitoring program was defined and it will be implemented gradually during the commissioning of the ESS facility.

References

[1] IAEA SRS19 IAEA Safety Report Series 19 Generic Models for Use in Assessing the Impact of Discharges of Radioactive Substances to the Environment, International Atomic Energy Agency Vienna 2001 ISBN 92–0–100501–6
[2] Ene D, 2016, Radiological and environmental impacts of the activated air inside the tunnel of ESS accelerator. Proceedings of SATIF13 workshop.
[3] Ene D, 2016, Source Term to the Environment from the ESS Target Station. Proceedings of SATIF13 workshop
[4] Ene D., 2017, ESS Environment Impact Assessment. Status report. Internal report ESS-0050117
[5] Ene D., 2016, Assessment of environmental consequences of the normal operations of ESS facility. Input data. Internal report ESS-0028551.
[6] Stenberg K. et al, 2017, Radionuclide Transport and Dose Calculations models for ESS. Internal report ESS-0052265
[7] Mokhov N. et al. 2017, Activation of soil around ESS tunnel. Internal report ESS-0098496
[8] Bugay D., 2016, Groundwater flow and radionuclide transport models for groundwater pathway analyses for ESS. Internal report ESS-0051622
[9] Bugay D., 2017, Groundwater pathway analyses for ESS: Radionuclide transport modeling results for the updated radioactivity source-term dataset. Internal report ESS-0114593

IOP Conf. Series: Journal of Physics: Conf. Series **1046** (2018) 012019 doi:10.1088/1742-6596/1046/1/012019

Leakage of radioactive materials from particle accelerator facilities by non-radiation disasters like fire and flooding and its environmental impacts

A Lee, N S Jung, L Mokhtari Oranj and H S Lee*

Pohang Accelerator Laboratory, POSTECH, Pohang, 37673, Republic of Korea

*Email: lee@postech.ac.kr

Abstract. The leakage of radioactive materials generated at particle accelerator facilities is one of the important issues in the view of radiation safety. In this study, fire and flooding at particle accelerator facilities were considered as the non-radiation disasters which result in the leakage of radioactive materials. To analyse the expected effects at each disaster, the case study on fired and flooded particle accelerator facilities was carried out with the property investigation of interesting materials presented in the accelerator tunnel and the activity estimation. Five major materials in the tunnel were investigated: dust, insulators, concrete, metals and paints. The activation levels on the concerned materials were calculated using several Monte Carlo codes (MCNPX 2.7+SP-FISPACT 2007, FLUKA 2011.4c and PHITS 2.64+DCHAIN-SP 2001). The impact weight to environment was estimated for the different beam particles (electron, proton, carbon and uranium) and the different beam energies (100, 430, 600 and 1000 MeV/nucleon). With the consideration of the leakage path of radioactive materials due to fire and flooding, the activation level of selected materials, and the impacts to the environment were evaluated. In the case of flooding, dust, concrete and metal were found as a considerable object. In the case of fire event, dust, insulator and paint were the major concerns. As expected, the influence of normal fire and flooding at electron accelerator facilities would be relatively low for both cases.

1. Introduction

In Korea, particle accelerator facilities are classified as radiation generators (RG) in accordance with Nuclear Safety Act. Therefore, to satisfy the location standard of the radiation generator use and storage facility, the particle accelerator facility should be installed at a place with the less risk of a fire, inundation or ground subsidence [1]. There are also design criteria for reactor facilities with detailed requirements related to fire and flooding. In applying the current regulation, small radiation generators do not pose a serious problem because of the limited parts and areas where radiation can be generated. However, in order to apply the criteria to the licensing process of large accelerator facilities, it is required to adopt a somewhat excessive law that is not related to the degree of radioactivity of each facility. For this situation, detailed and reasonable requirements that can be used by regulators, builders and operators of large accelerator facilities are needed.

In this study, fire and flooding at particle accelerator facilities were considered as the non-radiation disasters which result in the leakage or spread-out of radioactive materials. To analyse the expected effects due to each disaster, the case study on fired and flooded particle accelerator facilities was carried out with the property investigation of interesting materials presented in the accelerator tunnel and the activity estimation. Five major materials in the tunnel were considered: dust, insulators, concrete, metals and paints. Additionally, the activation levels on the concerned materials were calculated using three different Monte Carlo codes: MCNPX 2.7+SP-FISPACT 2007, FLUKA 2011.4c and PHITS 2.64+DCHAIN-SP 2001. Those estimations were carried out and compared each other for the different beam particles (electron, proton, carbon and uranium) and the different beam energies (100, 430, 600 and 1000 MeV/nucleon). This work aimed to evaluate the impacts to the environment, considering the leakage path of radioactive materials due to fire and flooding, and the activation level of selected materials.

2. Materials and methods

2.1. Case study of fire and flooding in particle accelerator facilities

2.1.1. Fire. Seven cases of fire accidents occurred in particle accelerator facilities were summarized in table 1. The fire accident that occurred in the underground accelerator room of Moscow nuclear institute, which had not been reported for its causes. All of other fire cases occurred on the electrical equipment. For J-PARC (Japan Proton Accelerator Research Complex) fire case, there was no leakage of radioactive material, taking into consideration the incident region of fire..

Table 1. List of fire case in particle accelerator facilities.

Facilities (Nation)	Cause	Incident region	Progress and action	Leakage of radioactive material
Fermilab (U.S.) [2]	Over-current in a cable	34 wire ribbon cable	- Date : Oct. 3, 1987 - Fire spread along cable trays to eventually involve 4 particle detectors. - 5 sprinklers were activated.	Not described
Cornell's Particle Accelerator (U.S.) [3-5]	Power supply equipment used for 30 years	Wilson Synchrotron Laboratory	- Date : Sep. 16, Oct. 14, 2009 - Fire was quickly put out. - The laboratory is located in the north section of the underground accelerator ring	Not described
Moscow nuclear research centre (Russia) [6,7]	-	Alikhanov Institute of Theoretical and Experimental Physics	- Date : Feb. 5, 2012 - A non-operational 60-year-old reactor and the collider shut off on Dec. 25 were located. - No radiation sources in collider area. - No flames and only smoke came from an area with power cables and could not affect any nuclear materials.	Not described
SLAC (U.S.) [8-11]	Magnet	Beam switching yard	- Date : 2013	Not described
	Electrical fire from switchgear cabinet	Sector 26 of the klystron gallery	- Date : Jun. 25, 2014 - Power to the switchgear was de-energized by SLAC high voltage electricians. - Fire was extinguished in 45 minutes.	Not described
J-PARC (Japan) [12-14]	Diesel generator	-	- Date : Jul. 2014	No
	Circuit design of newly installed transformer	Muon D-line area in the 2nd Experimental Hall	- Date : Jan. 16, 2015 - The place is located in the Materials and Life Science Experimental Facility - Fire was immediately extinguished with a fire extinguisher - No radiation exposures and no injuries	No

2.1.2. Flooding (Inundation). Flooding events at Fermilab and J-PARC facilities were investigated and table 2 shows the details. Fermilab has a pond where the rainwater is collected and used as cooling water. When a large amount of rainfall occurred, Fermilab was flooded because of the pond [15,16]. Two cases of flooding that occurred in J-PARC were investigated. The earthquake that struck Japan in March 2011 cut off the power supply to the pump used to discharge the groundwater in the accelerator tunnel, and the linear accelerator tunnel was flooded. It should be noted that the groundwater infiltrated by concrete during the weekly immersion period was measured as a strong alkaline (pH 11) at the center of linear accelerator tunnel. This caused the aluminium pre-amplifier box located at the bottom of the tunnel to be corroded. Therefore, not only water but also alkali solvents were considered in the flooded situation. The flooding that occurred in 2014 was caused by the problem with the valve connection in the leak test for the new cooling water line (CW line). After the incident, radiation survey were carried out but no leakage of radioactive material was observed [18-22].

Table 2. List of flooded particle accelerator facilities.

Facilities	Cause	Incident region	Progress and action	Leakage of Radioactive material
Fermilab (U.S.) [15,16]	Rainfall (46.72 cm)	Main ring tunnel (Linac, booster, anti-proton source: not flooded)	- Date : Jul. 17-18, 1996 - Tevatron accelerator was shut down. - Water was pumped on Jul. 18 - Recovery efforts at the laboratory were under way by Jul. 22.	Not described
	Rainfall	-	- Date : Apr. 18, 2013 - Closed for 17 hours. - Damage repair estimates were done and electrical equipment were repaired and replaced.	Not described
J-PARC (Japan) [17-22]	Loss of power for water pumping system	Inside of accelerator tunnel	- Date : Later than Mar. 11, 2011 - 1 cm deep water was on the floor (Mar. 17) and Max. depth of water reached 10 cm (Mar. 24) - pH 11 of water in the tunnel. - Aluminum pre-amplifier boxes were corroded. - Radiation survey work at the RCS and MR tunnel were carried out.	No
	Mis-connected valve.	Muon facility	- Date : Sep. 2014	-

2.2. Selection of interesting material inside the particle accelerator tunnel
The materials that could leak to the outside of the accelerator tunnel when fire or flooding occurred inside the accelerator tunnel were selected in the case study. Five major materials in the tunnel were investigated: dust, insulators, concrete, metals and paints.

- Dust normally exists on the accelerator devices and the tunnel wall. According to the IAEA report (Technical Reports Series No. 188), exposure due to activated dust can be done when an operator enters a tunnel after a shutdown of the accelerator and an evaluation of dust should be performed in such a situation. [34].
- Concrete is the most basic material that constitutes the accelerator tunnel, and plays an important role as a structure safety and radiation shield [35].

- Many kinds of metals are commonly used as constituent of accelerator devices. Residual radiation from metal devices is a major contributor to exposure to maintenance worker. Four types of metals were considered: aluminium, iron, stainless steel, and copper.
- Carbon compounds are often used in the accelerator tunnels as insulators like control and signal processing cables and wires. Three types of carbon compounds were considered in this study: polyvinylchloride (PVC), polytetrafluoroethylene (Teflon), and polyimide (Kapton).
- A few kinds of paints coat the accelerator tunnel wall, floor and components. The concerned kinds of paint were acrylic resin-based paints used for coating accelerator tunnel walls and epoxy-based paints used for the accelerator tunnel flooring and metal parts painting. The paint type used for Pohang Accelerator Laboratory was investigated [36].

2.3. *Activation calculation of the material inside the accelerator tunnel*

The activation levels of the concerned materials were calculated using several Monte Carlo codes, FLUKA 2011.4c [37], MCNPX 2.7 [38] +SP-FISPACT 2007 [39] and PHITS 2.64 [40] +DCHAIN-SP 2001 [41]. In this calculations, several conditions of thebeam particle, the target condition and the geometry were analyzed. The activation levels of each material were estimated in the cases of different beam energies of 100, 430, 600 and 1,000 MeV/nucleon and different beam particles of electron, proton, carbon and uranium. The shape of all particle beams was assumed to be flatwith 1 cm diameter, and the target was a cylindrical copper with a radius of 2 inches (= 5.08 cm) and a thickness of 20 cm. The target was surrounded by disk-shaped samples with a volume of 10 cm^3 (diameter: 2.52 cm, thickness: 0.5 cm), shown as figure 1, and the materials of activation samples were ST-37 and ST304L for metals, ANSI-ANS concrete for concrete, and Kapton for insulators. For paint, it is assumed to be epoxy resin based on the fact that the paint used for the tunnel floor and metal painting is epoxy resin type.

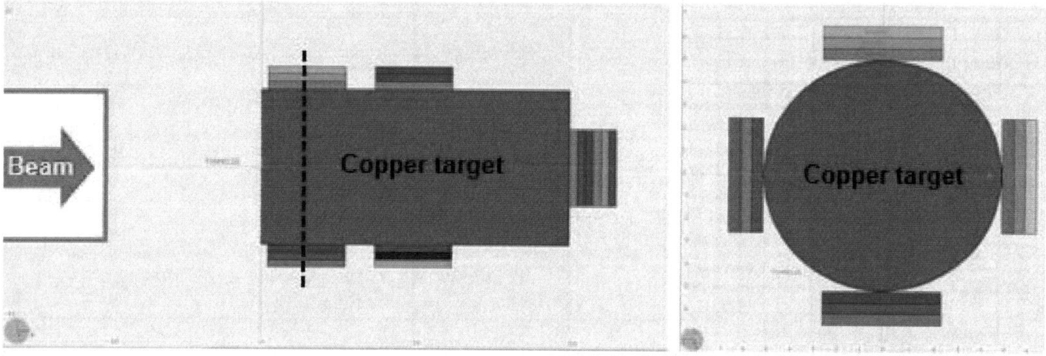

Figure 1. Schematic view of the target and samples-cross-sectional view (left) zy axis. (right) xy axis

3. Results and discussion

3.1. *Characteristics of materials to be considered in case of fire and flooding at the tunnel*

3.1.1. For fired particle accelerator tunnel. According to the case study, electric fire is probable by a short circuit, and an overcurrent, etc. and the carbon-compound based insulator was assumed as an ignition source. The characteristics of three insulators (PVC, Teflon and Kapton) were investigated and described in table 3. The flash point and the spontaneous ignition temperature measured according to the ASTM D1929 standard of material test and material specifications were used [42-44] at this study. There are two situations that can cause a fire. The first is when the temperature of the insulator reaches

the flash point by the heat. And the second is when spontaneous ignition may occur because it reaches the ignition point.
Considering the characteristics of the insulator, the insulator will be the material difficult to be burnt in general circumstances. However, in other studies, a combustion in tunnel were reported at situations which a large amount of insulators and cables are placed in the tunnel. The temperature inside the tunnel when a fire occurs due to the combustion of the insulator inside the accelerator tunnel is good reference to consider the combustion of the cable placed in the cable tray. The maximum temperatures measured by the thermocouples which were arrayed at a distance of 1.5 m from the cable tray were about 150 °C for a 0 mm spacing of the cable and 200 ~ 250 °C for a 9.2 mm spacing [45]. The experimental results on combustion temperature of the cable were also found in a study on the functional failure temperature of cables in a nuclear power plants. As a result of the cable combustion test, it was confirmed that most of the cables start to be burned at 450 °C of tunnel fire temperature. [46]. Based on the characteristics of the cable and the experimental results of these studies, it is assumed that insulator was burned at 450 °C and the temperature inside the tunnel that can be reached through the combustion of the insulator is 150 ~ 250 °C.

Table 3. Properties related with fire of insulators (PVC, Teflon, and Kapton).

	Maximum service temperature, Air (℃)	Flammability, UL94 [a] [68,69]	Melt temperature (℃)	Flash-ignition [a] (℃) [42-44]	Self-ignition [a] (℃) [42-44]
Polyvinylchloride (PVC)	40~100	HB[b] ~ 5VA[c]	154~213	330~400	385~450
Teflon (PTFE)	482	V-0[d]	-	560	580
Kapton (polyimide)	400	V-0	-	-	-

[a] Flammability, UL94 : "The Standard for Flammability of Plastic Materials for Parts in Devices and Appliances", is one of the most widely accepted flammability performance standards for plastic materials..
[b] HB : slow burning on a horizontal specimen less than 76 mm·min^{-1} for thickness less than 3mm.
[c] 5VA : After flame or afterglow time ≤60 seconds after 5th flame application, specimen may not have a burn-through
[d] V-0 : burning stops within 10 seconds, no drips allowed.

The effect of fire on the concrete was investigated in terms of temperature change and summarized in table 4. Considering the temperature at which the explosive fracture and melting occurs (1,200 ℃) and the temperature at which the insulator can ignite (330 ~ 580 ℃), the possibility of explosion and melting of concrete is low at the particle accelerator tunnel. In addition, concrete is a nonflammable material, so the possibility of leakage of radioactive material from concrete when a fire occurs is considered to be relatively low [47].

Table 4. Property-changes of concrete depending on temperature [47].

	Property-changes
100℃ or higher	- Free pore water release
100~200℃	- Physically adsorbed water is released and concrete is shrunk. - Safety is ensured.
300℃ or higher	- Cement hydrate in the concrete chemically altered
400℃ or higher	- Release chemical combined water

500℃ or higher	- Calcium hydroxide($Ca(OH)_2$) is thermally decomposed (pyrolysis), durability significantly reduced and 50% reduction in strength by heating
600~800℃	- Cement paste is shrunk and aggregate has blowout. - 80% reduction in strength by heating
1,000~1,200℃	- Explosive fracture (Spalling failure) occurs
1,200℃	- Melting occurs

The characteristics of the metal were investigated in terms of melting point, boiling point and vapor pressure shown in table 5 [48-51]. These temperatures are much higher than the temperature that can be reached by the combustion of the insulator inside the tunnel. It is expected that the possibility of burning and the effect of fire on metal in the event of fire will be significantly low and the leakage of radioactive material rarely occurs.

Table 5. Properties of metals (aluminium, steel, stainless-steel, and copper)

	Melting Point (℃)	Boiling Point (℃)	Vapor pressure (°C)
Aluminum [48]	660.37	2,519	2,517 (100 kPa)
Steel (iron) [49]	1,535	2,861	2,859 (100 kPa)
Stainless-steel [50]	1,400~1,455	2,750	-
Copper [51]	1,083.2~1,083.6	2,562	2,563 (100 kPa)

The paint is assumed to be ignited by the contact of flames generated on insulator. The flash point is mainly considered. Both acrylic resin-based paints and epoxy-based paints have a flash point of about 30 °C and the conditions are summarized in table 6 [52-63].

Table 6. Fire properties of paints (acrylic resin-based paint and epoxy resin-based paint).

	Coating place	Flash-ignition (℃)	Self-ignition (°C)
Acrylic resin-based paint - ACRYDIC A-846 [52]	Accelerator tunnel wall	36	480
Epoxy resin-based paint - Epocoat210 : floor	Accelerator tunnel floor	24	-
- NUKOPOX PRIMER EP103, NUKOPOX TOPCOAT ET597 : metal component [53-63]	Metal component (magnet, support)	EP103 : 25~26 ET597 : 23~25	EP103 : 343~464 ET597 : 343~464

3.1.2. For flooding particle accelerator tunnel. In the case of flooding, the materials placed inside the tunnels were analysed with respect to the solubility and corrosiveness to water and alkali solvents, and to water absorption rate and reactions with water. Dust on the tunnel wall and accelerator devices would be easily rinsed off with water when the flooding occurs inside the accelerator tunnel. Concrete can react with water and the hydration reaction occurs. Among the compounds of the cement, C_2S reacts with water to form calcium hydroxide ($Ca(OH)_2$) and calcium silicate hydrate (C-S-H), and then the liquid phase becomes highly alkaline [64]. This phenomenon was confirmed in the J-PARC accelerator tunnel flooding case. The solubility and corrosiveness of metals in water and alkali solvents were investigated and summarized in table 7 [48-51]. Four types of metals are not dissolved or corroded in water, but are soluble in alkaline solvents. So the immersion is important issue to be considered.

Table 7. Solubility for Water and Alkaline of Metals (aluminum, steel, stainless-steel, and copper).

	Solubility (Corrosiveness)	
	Water	Alkaline
Aluminum [48]	Insoluble	Dissolved
Steel (iron) [49]	Insoluble	Dissolved
Stainless-steel [50]	Insoluble	Dissolved
Copper [51]	Insoluble	Dissolved (slowly)

In the case of insulators, the water absorption, the resistance to water and alkali solvents, and insolubility were considered and summarized in table 8 [32, 65-67]. The water absorption rate of the insulators was used to determine the possibility that the radionuclides produced in the insulators were leaked to the outside through the absorption and re-emission of water. The water absorption rates of three kinds of insulators (PVC, Teflon, Kapton) were found to be as low as 0 ~ 1.3% [32]. PVC and Teflon have low solubility and good chemical resistance to strong alkaline solvent. It is likely to dissolve in a flooded environment because of its good resistance to most cast of acids and alkalis.

Table 8. Absorption and resistance for water and alkaline of insulators (Polyvinylchloride, Teflon and Kapton) [32,33-46].

	Water Absorption (%) [32]	Resistance		Solubility
		Water	Alkaline	
Polyvinylchloride (PVC) [65]	0.06	Good resistance	Good resistance	Insoluble
Teflon (PTFE) [66]	0 (<0.01)	Excellent chemical resistance	Excellent chemical resistance	Insoluble
Kapton (polyimide) [67]	1.3	Greater resistance	-	Insoluble

Paint was classified into the acrylic resin type paint and the epoxy type paint. Characteristics of each type of paints were investigated and this is shown in table 9. Acrylic resin-based paints (Aekyung Chemical, ACRYDIC A-846) have excellent chemical resistance against 5% NaOH and 5% H_2SO_4, respectively [52]. Epoxy-based paint (Epocoat 210 from Samhwa paint) is used for coating the bottom of accelerator tunnel. This paint has excellent adhesion and durability for water and alkali solvents, as well as excellent abrasion resistance, rust prevention, chemical resistance, water resistance, water insolubility, water resistance, acid resistance and alkali resistance [53]. Paint used for coating the electronic devices (KCC NUKOPOX PRIMER EP103, NUKOPOX TOPCOAT ET597) has excellent water resistance, acid resistance and alkali resistance. Therefore, paint is less likely to react with water and dissolve, so it is unlikely that radioactive materials will leak out in flood situations [52-63].

Table 9. Properties of paints (acrylic resin-based paint and epoxy resin-based paint).

	Coating place	Resistance		Solubility
		Water	Alkaline	
Acrylic resin-based paint - ACRYDIC A-846 [52]	Accelerator tunnel wall	-	Resistant (Excellent for 5% NaOH)	-

Epoxy resin-based paint - Epocoat210 - NUKOPOX PRIMER EP103 [48] - NUKOPOX TOPCOAT ET597 [53-63]	Accelerator tunnel floor	Resistant	Excellent resistance	Insoluble (water)
	Metal component (magnet, support)	Excellent resistance	Excellent resistance	-

3.1.3. Flow of influence on material during fire-flooding event. The influences on the material were designed in various cases based on the characteristics of the materials in the event of fired or flooded particle accelerator tunnel, shown in figure 2 and 3. First, it is assumed that fire in the accelerator tunnel is an electric fire caused by leakage of current, overcurrent (overload), etc., and that ignition occurs on the insulator. In addition, due to the ignition of the insulator, the paint on the accelerator device will also burn, which is expected to raise the temperature of the surrounding materials. Concrete is a non-flammable material so will not be directly combusted. With the consideration on the property change due to temperature rise, it is expected that the effect of fire on concrete will be relatively low since the tunnel temperature will be considerably lower than the temperature at which explosion or melting occurs.

Considering that the melting point, boiling point and vapor pressure of metals are mostly over 2,000 ℃, the possibility of leakage of radioactive materials is expected to be ignorable. Dust will not burn significantly since there is generally a small amount of dust around the cable. However, there is a possibility of leaking into the outside environment due to activated dust floating through air. If there is a large amount of dust accumulated for several years, accidents such as ignition and dust explosion may occur, so that dust management should be performed regularly.

In the case of flooded accelerator tunnel, dust located near the floor and devices can be easily washed away by water, hence consideration of the amount of dust and the activation level are needed. Concrete is supposed to produce alkali solvent by the hydration reaction and this can affect other materials. But additional consideration is needed, such as the immersion duration and hydration reaction speed. Metals are expected to dissolve and corrode by alkali solvent, but this will be determined according to flood (water) level. According to the case study, the water level was below the girder height in the general flood situation and the accelerating devices were not flooded. Insulators and paints are less likely to leak radioactive materials due to their reaction with water, according to the properties of Teflon and PVC and paint. However, the degradation of insulators caused by radiation damage is also a factor to be considered.

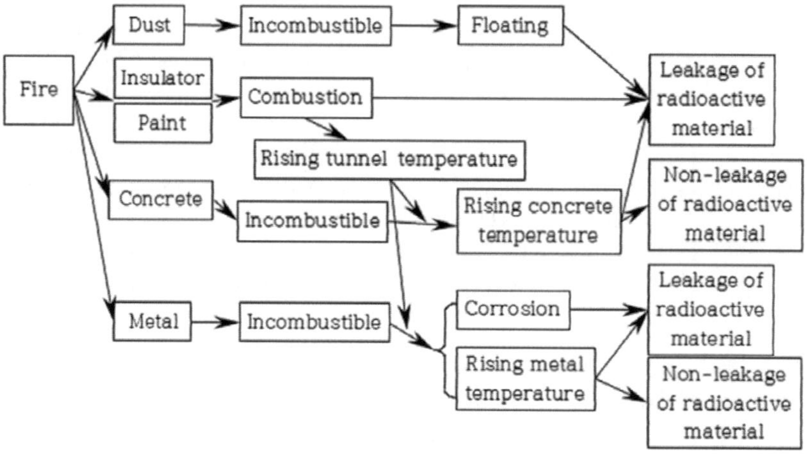

Figure 2. Assumed flow of affection on materials and leakage of radioactive material.

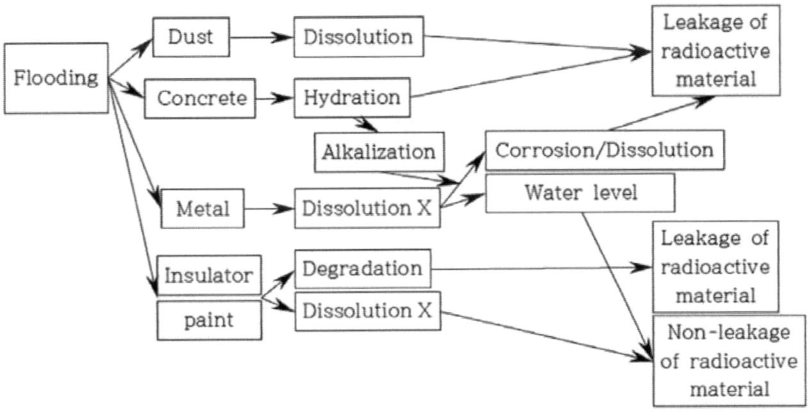

Figure 3. Assumed flow of affection on materials and leakage of radioactive material.

3.2. Calculation result of the interesting material inside the accelerator tunnel

3.2.1. Comparison of Monte Carlo codes and activated materials. Figure 4 shows the specific activities generated in the metal (ST304L) calculated by Monte Carlo codes. For the proton beam, the calculation results of the three Monte Carlo codes were similar, but the results of MCNPX code were the largest. However, the calculation results of FLUKA code were the highest for heavy ion beams (carbon and uranium). The radionuclides generated in the disk-shaped metal (ST304L), concrete (ANSI-ANS), insulator (Kapton) and paint (epoxy) around the target when the carbon beam of 430 MeV energy enters the copper target are shown in table 10. Radionuclides generated in each material after 1 day cooling time were analysed to reduce the effects on the short-lived radionuclides (less than 1 day). ST304 contains 69.5% of ^{26}Fe and 17.8% of ^{24}Cr as well as other impurities such as ^{6}C and ^{14}Si. As result of activation calculation, $^{44,\,44m,\,47}$Sc, ^{48,49}V, ^{51}Cr, ^{52,54}Mn, ^{55}Fe, ^{58,58m}Co and ^{57}Ni were mainly produced. In ANSI-ANS concrete, ^{3}H, ^{7}Be, ^{22,24}Na, ^{37}Ar, ^{54}Mn, ^{55,59}Fe and ^{57,60}Co were produced and most of the radionuclides are long-lived (several hundred to several decades). For Kapton and Epoxy, the main radionuclides produced are ^{3}H, ^{7,10}Be, and ^{11,14}C and can contribute to residual radiation as a long-lived radionuclides. In addition, the ratio between calculated specific activities of each radionuclides in interested materials are shown in figure 5. As a result, FLUKA code produced more radionuclides than MCNPX and PHITS code.

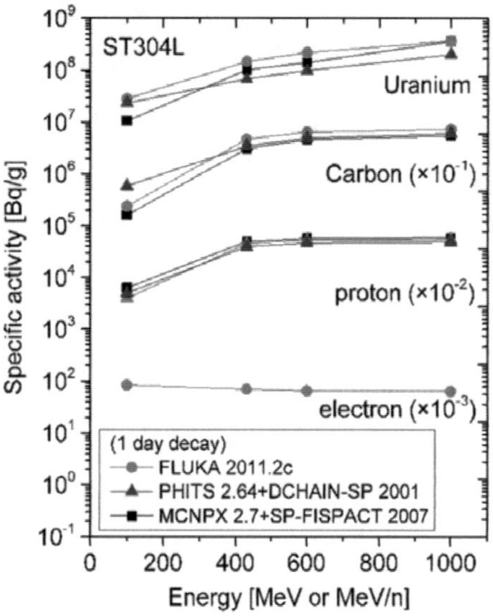

Figure 4. Total specific activities calculated by three Monte Carlo codes.

Table 10. Specific activity of major materials in accelerator tunnel (cooling time: 1 day).

Material	Dominant radionuclides from FLUKA code calculation
Metal (ST304L)	$^{44,44m,47}Sc$, $^{48,49}V$, ^{51}Cr, $^{52,54}Mn$, ^{55}Fe, $^{58,58m}Co$, ^{57}Ni
Concrete (ANSI-ANS)	^{7}Be, ^{24}Na, ^{31}Si, ^{32}P, ^{37}Ar, $^{42,43}K$, $^{45,47}Ca$, ^{51}Cr, $^{55,59}Fe$
Insulator (Kapton)	^{3}H, $^{7,10}Be$, $^{11,14}C$
Paint (Epoxy)	^{3}H, $^{7,10}Be$, $^{11,14}C$

IOP Conf. Series: Journal of Physics: Conf. Series **1046** (2018) 012019 doi:10.1088/1742-6596/1046/1/012019

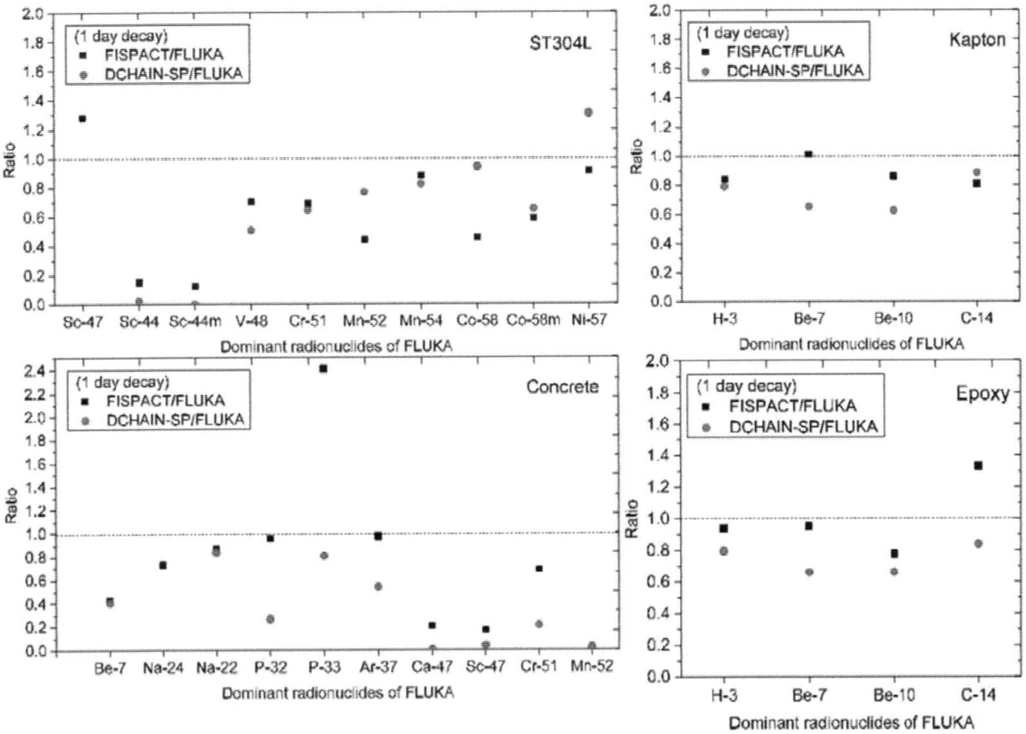

Figure 5. Ratio between calculated specific activities of each radionuclides in interested materials irradiated by 430 MeV·n^{-1} ^{12}C using different Monte Carlo codes.

The specific activities in terms of arbitrary unit according to the cooling time was calculated and shown in figure 6. The specific activities of the interested materials decreased in the form of stairs and decreased by a factor of more than 10 times in 24 hours after the irradiation.

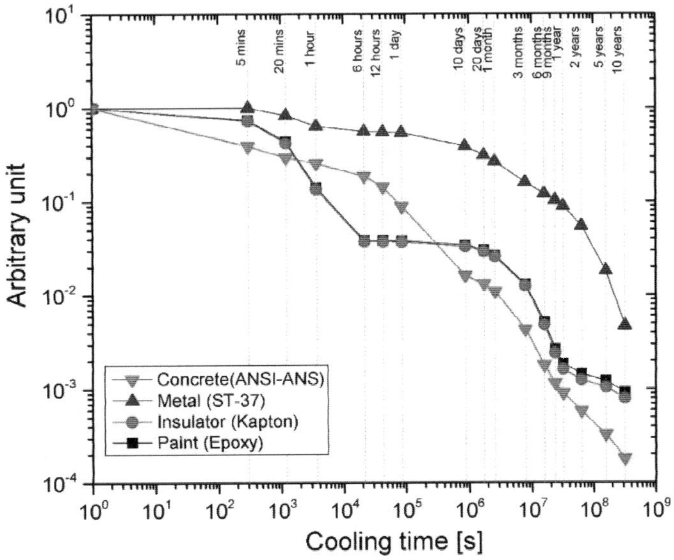

Figure 6. Specific activity of each radionuclide produced in insulator (Kapton) irradiated by 430 MeV·n-1 ^{12}C.

3.2.2. Comparison of the specific activities for the beam energy and particles. Figure 7 shows the clearance index of interested materials for various beam energies and particles. The higher beam energy and the atomic number of the beam particles, the higher specific activities and the higher ratio to the clearance index were calculated. Total specific activity of interested material was high in the order of uranium, carbon, proton and electron beam. The highest specific activity of ST-37 was calculated for uranium beam and is 5,200 times higher than for electron beam. In the case of the same incident beam, the specific activity is high in the order of metal, concrete, insulator and paint, and calculated to be 800 times higher than the clearance level. Therefore, it is necessary to consider the activation level for the different beam energy, particle and the material inside the tunnel.

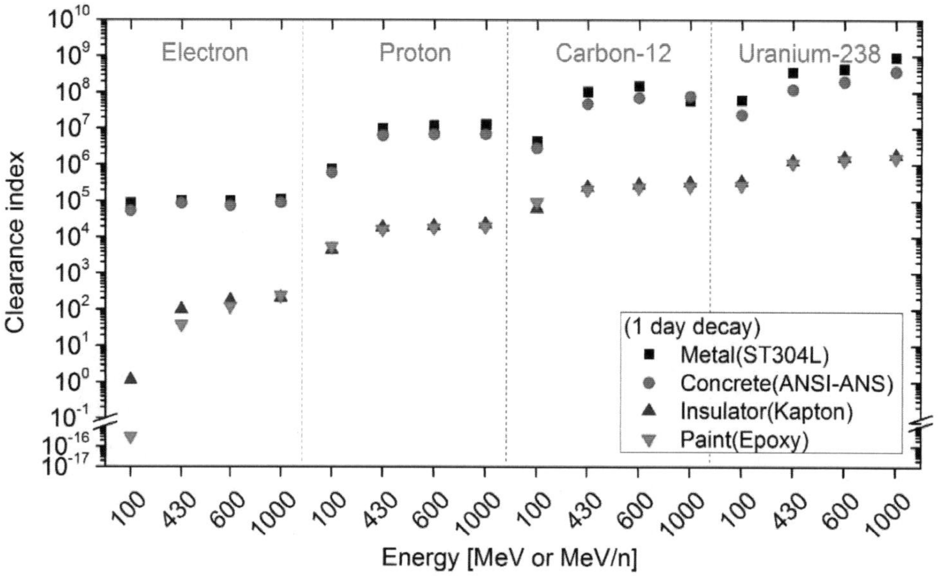

Figure 7. Clearance index of interested materials for different beam energy and particle.

3.3. Comparison of impacts on external environments in fired and flooded particle accelerator facilities

According to the assumed influence on materials, it was found that three materials can be affected by fire, which are dust, insulation and paint. Dust is expected to be negligible in general situations. However, it will be necessary to continuously manage the emission of dust in a region where dust can be accumulated and the beam loss region. Insulators and paints can be burned, but the activation level is significantly lower than those of other materials. For flooding, it is assumed that dust, concrete, and metal are likely to be affected by inundation. However, the amount of dust accumulation can be controlled by cleaning, and the activation level is low. It was expected that the exposure by activated dust and the influence on the external environment will be insignificant. Concrete is expected to elute alkaline ions by the reaction with water, and there is a possibility that radioactive nuclides will escape from this reaction. However, in the case of flooding occurred at the J-PARC facility, radioactive nuclides were not detected in the water. Metal is one of the main materials to be activated in the tunnel and activation level is 4 ~ 5 times higher than insulator according to the activation calculation. Metal can be corroded by alkaline water, however, it is considered that the metal device are placed on the girder so that the flooded water level must be evaluated together. Following the case study on flooding, the flooded water level was limited to below the girder so the possibility of leakage of radioactive material due to flooding of radioactive metal parts is expected to be low. In the case of insulator, it is assumed that the insulator needs to be considered depending on the degree of degradation due to activation.

4. Conclusion

In this study, fire and flooding at particle accelerator facilities were considered as the situation which results in the leakage or spread-out of radioactive materials. To analyze the possibility of leakage of radioactive materials, case studies on fired and flooded particle accelerator facilities, the property investigations of interesting materials, and the activity estimation were performed. In the case of fire, dust, insulator, and paint can burn and float into the air, so there is a possibility of leakage of radioactive material. However, considering the activation level of the interested material, the influence on the external environment is confirmed to be low. In the case of flooding, there is a possibility of leaking of radioactive material presented in dust and metal. If considering the amount of dust in the general situation, the influence on the external environment may be small. Metals are highly radioactive but additional considerations are needed to ensure that the level of flooded water to reach the activated metal devices. Only the alkaline water due to the reaction of water with concrete can be considered seriously.

The activation calculations for concrete, metal, insulator and paint were performed for the various conditions of different beam energies and particles. When the same beam energy and beam particles were irradiated to each material, the activation level was different depending on the materials and high in the order of metal, concrete, insulator and paint. When irradiated with the same beam energy, different kinds of incident beam particles, the activation level was high in the order of uranium, carbon, proton, and electron. Therefore, to develop more practical regulation, it is necessary to consider the difference depending on the beam energy, beam particle, and the area where activation is produced in the accelerator facility.

References

[1] Korea Institute of Nuclear Safety 2014 *Regulations on technical standards for radiation safety control, Etc*
[2] Hinds S S 2011 Fire lessons learned in the nuclear industry. American Glovebox Society
[3] http://cornellsun.com/blog/2009/09/17/electrical-unit-catches-fire-at-synchrotron-laboratory/
[4] http://cornellsun.com/blog/2009/10/14/old-equipment-sparks-fire-at-synchrotron/
[5] http://bilmes.blogspot.kr/2009/09/fire-in-cornells-particle-accelerator.html
[6] http://www.reuters.com/article/2012/02/05/us-russia-fire-nuclear-idUSTRE8140J520120205
[7] http://bellona.org/news/nuclear-issues/accidents-and-incidents/2012-02-update-fire-breaks-out-at-central-moscow-nuclear-research-institute
[8] http://news.stanford.edu/news/2014/june/slac-fire-story-062714.html
[9] http://www.dailymail.co.uk/wires/ap/article-2670554/Fire-shuts-Stanford-linear-accelerator.html
[10] http://www.almanacnews.com/news/2014/06/27/menlo-park-fire-temporarily-shuts-down-slac-linear-accelerator
[11] https://news.slac.stanford.edu/announcement/electrical-fire-sector-26
[12] https://j-parc.jp/en/topics/20150219press.html
[13] http://j-parc.jp/en/topics/20150116press.html
[14] Saito N 2015 Overview of recovery efforts from the incident. *The 2nd Int. Symp. on Safety in Accelerator Facilities* (Tokai)
[15] Fermi National Accelerator Laboratory 1996 *FermiNews* **19** pp 1-4
[16] U.S. Department of Energy 2013 *Annual Laboratory Planning Fermilab* **22**
[17] MIURA A 2011 Progress of beam instrumentation in J-PARC linac. *Int. Beam Instrumentation Conf.* (Ibaraki)
[18] Yamamoto K 2012 J-PARC recovery status. Institute of High Energy Physics
[19] Hasegawa K 2012 Recovery of the J-PARC linac from the earthquake. *Linear Accelerator Conf.* (Tel Aviv)
[20] Hasegawa K, Kinsho M and Koseki T 2011 Status of J-PARC accelerator facilities after the great east Japan earthquake *Int. Particle Accelerator Conf.* (San Sebastian)

[21] Yamamoto K 2012 Comparison of the residual doses before and after resumption of user operation in J-PARC RCS. *Int. Particle Accelerator Conf.* (New Orleans Louisiana)
[22] Oguri H, Ikegamia K, Ohkoshia K, Koizumia I, Namekawab Y, Uenoa A, Takagia A and Yamazakia S 2013 Operation status of the J-PARC negative hydrogen-ion source. *American Institute of Physics Conf.* 379-85
[23] Charalambus S T and Rindi A 1967 *Nuc. Instr. and Meth.* **56** 125-35
[24] U.S. Department of Energy 1979 *Radiological environmental impact of high-energy accelerators. LBL-6169* pp 82-3
[25] Pohang Accelerator Laboratory 1997 *Dust radioactivity in the tunnel of Pohang light source. PAL-PUB-97-017* pp 1-5
[26] GSI *Proposal for experiment S291* Residual radioactivity induced by U ions - experimental investigation and longtime predictions.
[27] GSI *Decontamination and surface properties of the controlled areas of FAIR* 2012
[28] Masumoto K, Toyoda A, Eda K, Izumi Y, Shibata T *J. Radioanal. Nucl. Chem.* 465-9
[29] Sonck M, Buls N, Hermanne A and Eggermontcea G 2000 Radiological and economic impact of decommissioning charged particle accelerators. *Waste Management Conf.* (Tucson)
[30] Fasso A, Silari M and Ulrici L 2000 *Nucl. Sci. Tech.* 827-34
[31] Eggermont G X, Buls N and Hermanne A 1996 *Decommissioning analysis of a university cyclotron*
[32] CERN European Organization for Nuclear Research 1979 *Compilation of radiation damage test data - Part1: Cable insulating materials* pp 5-8
[33] KCC. Heavy duty coating technical data sheets
[34] International Atomic Energy Agency 1979 *Radiological safety aspects of the operation of electron linear accelerators. Technical reports series no. 188* pp 131-6
[35] Korea Atomic Energy Research Institute 2008 *Proton engineering frontier project -Radiation shielding and safety design. KAERI/RR-2888/2007* pp 64-5
[36] Pohang Accelerator Laboratory 2011 *Painting specifications for linear accelerator and storage ring* pp 1-17
[37] Fasso A, Ferrari A, Ranft J and Sala P R 2005 *The FLUKA Code: a multi-particle transport code. CERN-2005-10, INFN/TC_05/11, SLAC-R-773*
[38] Los Alamos National Laboratory *2011 MCNPXTM user's manual version 2.7.0. LA-CP-11-00438* pp 5-56 - 65
[39] Forrest R A 2007 *FISPACT 2007 User Manual EASY 2010, UKAEA FUS* 534
[40] Niita K, Sato T, Iwase H, Nose H, Nakashima H and Sihver L 2006 *Radiat. Meas.* **41** 1080- 90
[41] Takata H and Kosako K 1999 *Development of the DCHAIN-SP Code for Analyzing Decay and Build-up Characteristics of Spallation Products, JAERI-DATA/Code 99-008*
[42] American Society for Testing and Materials International 2016 *Standard test method for determining ignition temperature of plastics D1929-16* pp 1-6
[43] Occupational Safety and Health Research Institute 2003 *A study on the characteristics of fire hazard for plastic materials* pp 11-23
[44] Poly Marketing. PVC and fire 2
[45] Xianjia H, Kun B, Xiaoshuang L, Jianxin Y and Xi J 2015 A model for predicting temperature produced by upward spreading cable fire under natural ventilation, Energy Procedia 66 177-80
[46] Kim D H and Lim H S 2011 *J. KOSOS* **26** 41-45
[47] Lee C H, The impact of fire on buildings 2-3
[48] https://ko.wikipedia.org/wiki/%EC%95%8C%EB%A3%A8%EB%AF%B8%EB%8A%84
[49] https://ko.wikipedia.org/wiki/%EC%B2%A0
[50] https://ko.wikipedia.org/wiki/%EC%8A%A4%ED%85%8C%EC%9D%B8%EB%A6%AC%EC4%EA%B0%95
[51] https://ko.wikipedia.org/wiki/%EA%B5%AC%EB%A6%AC

[52] Aekyung Chemical Co, LTD. ACRYDIC A-848-RN, ACRYDIC A-837/A-846 2003
[53] Samhwa Paints Industrial Co. LTD. Material safety data sheet(MSDS)-3100247 2014
[54] KCC. Material safety data sheet (MSDS) - EP103PTB 2013
[55] KCC. Material safety data sheet (MSDS) - EP103PTA-N.7.0 2013
[56] KCC. Material safety data sheet (MSDS) - ET597PTB 2013
[57] KCC. Material safety data sheet (MSDS) - ET597PTA-VQ63G 2013
[58] KCC. Material safety data sheet (MSDS) - ET597PTA-5Y7/6 2013
[59] KCC. Material safety data sheet (MSDS) - ET597PTA-2.5Y8.5/2 2013
[60] KCC. Material safety data sheet (MSDS) - ET597PTB 2013
[61] KCC. Material safety data sheet (MSDS) - EP103PTA-N.7.0 2013
[62] KCC. Material safety data sheet (MSDS) - EP103PTA-N.7.0 2013
[63] KCC. Material safety data sheet (MSDS) - ET597PTA-2.5G7/2 2014
[64] http://www.tycement.co.kr/kor/develop/data/cement_09.pdf
[65] Occupational Safety and Health Research Institute 2003 *A study on the characteristics of fire hazard for plastic materials* pp 11-23
[66] DuPont. *Properties handbook - Teflon PTFE. (7/96) 220313D* pp 1-34
[67] Ministry of Economy, Trade and Industry 1992 *High-tech industry technical dictionary. Gyeomjisa* p 158
[68] http://ulstandards.ul.com/standard/?id=94
[69] http://www.thermon.com/ThermoTips/Flammability%20Testing.pdf

IOP Conf. Series: Journal of Physics: Conf. Series **1046** (2018) 012020 doi:10.1088/1742-6596/1046/1/012020

Activation assessment of the soil around the ESS accelerator tunnel

I L Rakhno[1], N V Mokhov[1], I S Tropin[1], D Ene[2]

[1] Fermi National Accelerator Laboratory, Batavia, Illinois 60510, USA
[2] European Spallation Source, ESS-ERIC, Tunavagen 24, SE-221 00 Lund, Sweden

rakhno@fnal.gov

Abstract. Activation of the soil surrounding the ESS accelerator tunnel calculated by the MARS15 code is presented. A detailed composition of the soil, that comprises about 30 chemical elements, is considered. Spatial distributions of the produced activity are provided in both transverse and longitudinal directions. A realistic irradiation profile for the entire planned lifetime of the facility is used. The nuclear transmutation and decay of the produced radionuclides is calculated with the DeTra code which is a built-in tool for the MARS15 code. Radionuclide production by low-energy neutrons is calculated using the ENDF/B-VII evaluated nuclear data library. In order to estimate quality of this activation assessment, a comparison between calculated and measured activation of various foils in a similar radiation environment is presented.

1. Introduction

The ESS linear accelerator is planned to be operating at an average beam power of 5 MW for 40 years [1]. Various radionuclides will be produced in the surrounding soil and surface water by secondary particles (mostly neutrons) generated in the machine components and target by the proton beam with energy up to 2 GeV. Therefore, a detailed analysis is needed to predict residual activity of the soil and water outside the accelerator tunnel and target station. This study addresses distributions of activity of radionuclides produced in the soil around the ESS accelerator tunnel wall for two periods of time: one year and forty years of normal operation for a given operational scenario. Further spatial migration of the produced radionuclides in the soil is not considered in this paper and will be a subject of another dedicated study.

Only the contribution from the beam losses in the linac (from beginning to the end of the accelerator to the target region, up to a neutron shield wall) is considered here. The primary particles are not tracked beyond the rightmost shielding wall of the accelerator tunnel.

The detailed distributions of the produced radionuclides are provided in 30 spatial bins—3 longitudinal and 10 transverse bins—for a further migration analysis by a dedicated expert group. Total residual activity of the accelerator tunnel itself is calculated as well.

A brief description of computer codes used for the calculations is provided. In order to verify quality of the predicted residual activation, one compares calculated and measured residual activity of various foils irradiated in a similar radiation environment. The comparison reveals a good agreement between the measurements and calculations.

2. MARS15 and DeTra computer codes

IOP Conf. Series: Journal of Physics: Conf. Series **1046** (2018) 012020 doi:10.1088/1742-6596/1046/1/012020

2.1. MARS15 code
MARS15 is a general-purpose Monte Carlo code for modeling particle and heavy ion interactions with matter and charged particle transport in magnetic field in realistic accelerator three-dimensional structures [2, 3, 4]. The energy span covers many decades: from a multi-TeV region down to the thermal energies for neutrons and to keV region for other particles. The code has more than a 40-year history and for the last approximately 25 years is developed and maintained at Fermi National Accelerator Laboratory. In our energy region of interest—below 2 GeV—the following data libraries and nuclear collision models are used:

- Below 100 MeV (adjustable), TENDL-2015 library for protons, light ions and gammas, and CEM model (code) for all other projectiles;
- Mix-and-match procedure around 100 MeV between TENDL and CEM model (code);
- Exclusive modeling with CEM code for energies between 100 and 300 MeV;
- Mix-and-match procedure between CEM and LAQGSM model (code) between 300 and 500 MeV;
- Exclusive modeling with LAQGSM above 500 MeV.

Neutron interactions below 14 MeV are modeled using ENDF/B-VII library.
At present, DeTra code is implemented as a part of MARS15 code, and it can be invoked separately after MARS itself generates output files with calculated nuclide production rates in specified regions of interest.

2.2. DeTra code
DeTra code has approximately a 20-year history and was developed at Helsinki University and CERN for various calculations of residual activity at accelerators [5]. As the name suggests, the code performs decay, build-up and transmutation calculations by means of analytically solving Bateman equations for a given irradiation and cooling down scenario. The complexity of the decay chains and the number of nuclides are not limited as long as the nuclear data are known (supplied). At present, library NUDAT 2.6 is used as a source of nuclear decay data.

3. Geometry model of the ESS accelerator and tunnel with service buildings
The MARS15 computation model developed and described in detail in Refs. [6, 7] is used for normal operational beam losses for calculations presented in this paper. Only the major beam line components were modelled, and support structures and devices were excluded from the calculation model. The right-hand coordinate system origin (x=y=z=0) is at the beginning of the ESS linac, with x-axis pointing up, y-axis - to the right and z-axis along the beam in the linac horizontal section. In some cases, the s-axis coinciding with the beam axis is used. Figure 1 shows a vertical scan at y=0 through the ESS accelerator MARS15 model while a plan view at x=0 (beam level) is given in Fig. 2. A soil berm above the accelerator tunnel has a noticeable right (positive y) to left (negative y) slope with the highest vertical coordinate being x=9.3 m. The berm height is at x=8.42m above the beam (y=0), with the overall ground level at the site at x=2.1 m.
The radionuclide production is scored in transverse bins in soil immediately outside the concrete tunnel structures. The first five bins are 20-cm thick each, while the bins 6 thru 10 are 100-cm thick each. In addition, in order to provide more detailed data, the linac length was divided into three regions:

1. The low-beta region, with z from 0 to 176 m, with the corresponding maximum proton energy of 571 MeV;
2. the region downstream the region 1 and up to the end of the linac straight section, with z from 176 to 490.56 m, with the corresponding proton energies from 571 MeV to 1.991 GeV;
3. the region downstream the region 2 and up to the end of the accelerator, with z from 490.56 up to 580.5 m, with the proton energy of 1.991 GeV.

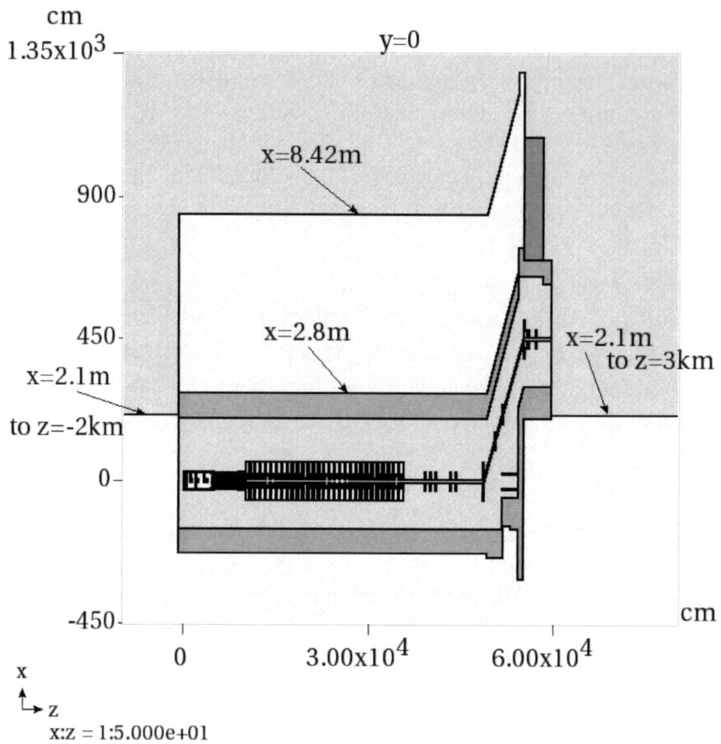

Figure 1. A side view of the ESS accelerator MARS15 model [6, 7].

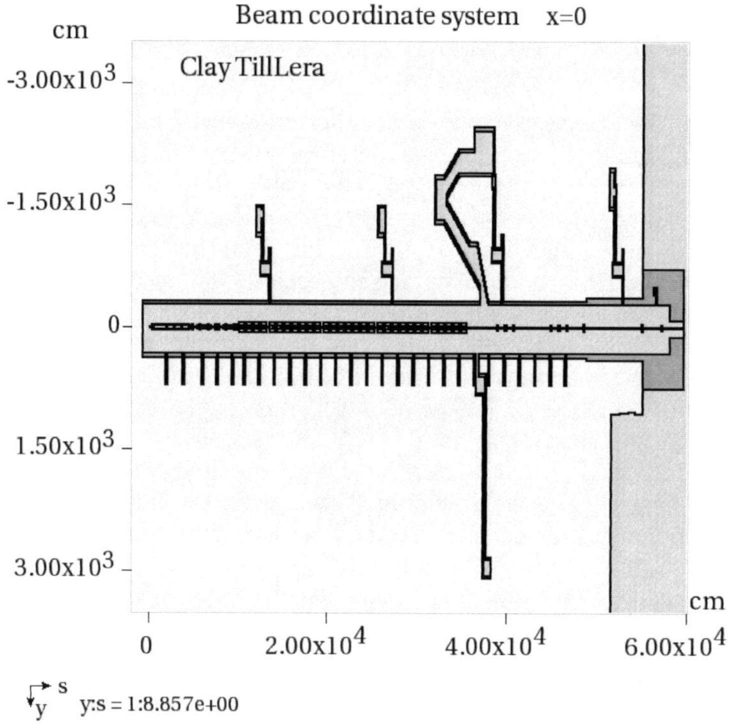

Figure 2. A plan view of the ESS accelerator and adjacent structures in MARS15 calculation model [6, 7]. Here the label "s" refers to the beam line coordinate.

IOP Conf. Series: Journal of Physics: Conf. Series **1046** (2018) 012020 doi:10.1088/1742-6596/1046/1/012020

Several fragments of the MARS15 model geometry, that are specific to this study on produced radionuclides, are given in Figs. 3-7. The various colours shown in the bins are used for technical reasons to specify volumes for scoring radionuclide production in the corresponding materials.

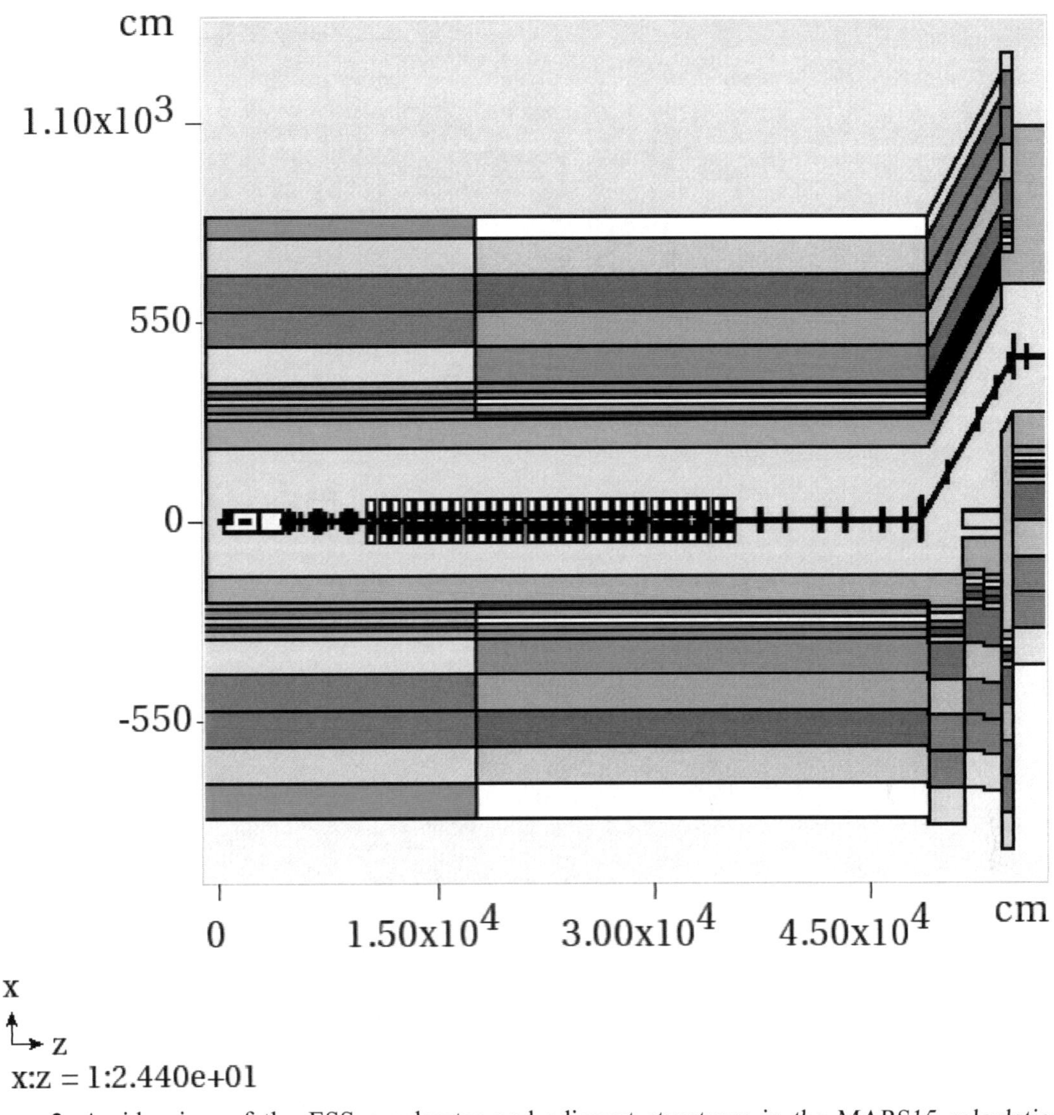

Figure 3. A side view of the ESS accelerator and adjacent structures in the MARS15 calculation model with transverse bins for scoring produced radionuclides.

IOP Conf. Series: Journal of Physics: Conf. Series **1046** (2018) 012020 doi:10.1088/1742-6596/1046/1/012020

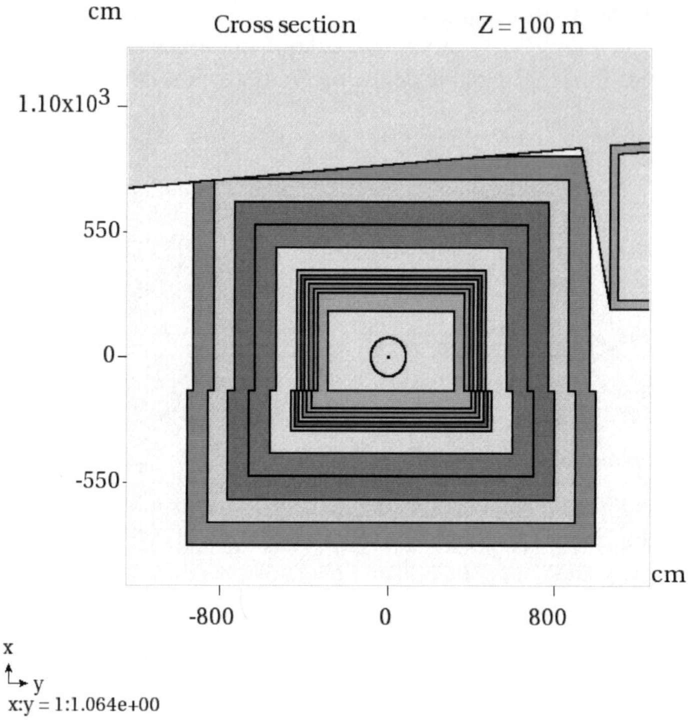

Figure 4. A cross section of the ESS accelerator and adjacent structures in the MARS15 calculation model in region 1 at z=100m with transverse bins for scoring produced radionuclides.

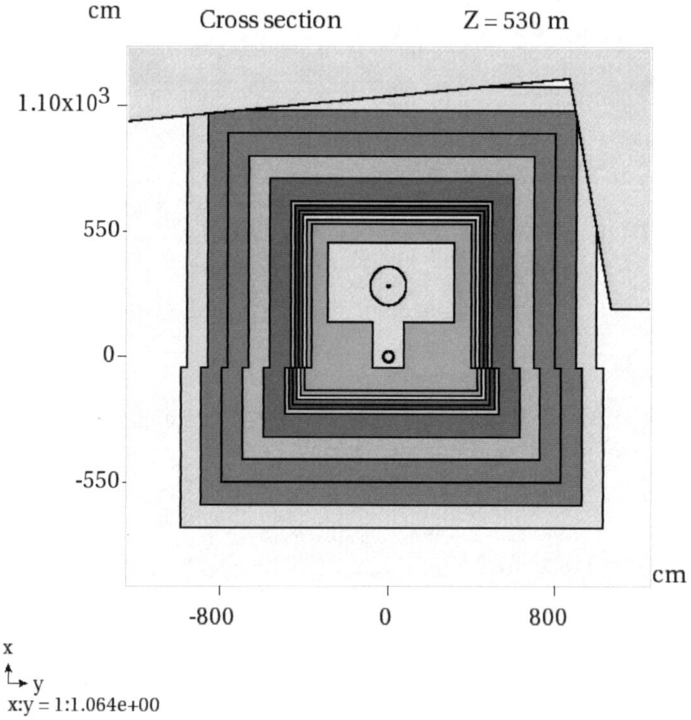

Figure 5. A cross section of the ESS accelerator and adjacent structures in the MARS15 calculation model in region 3 at z=530m with transverse bins for scoring produced radionuclides.

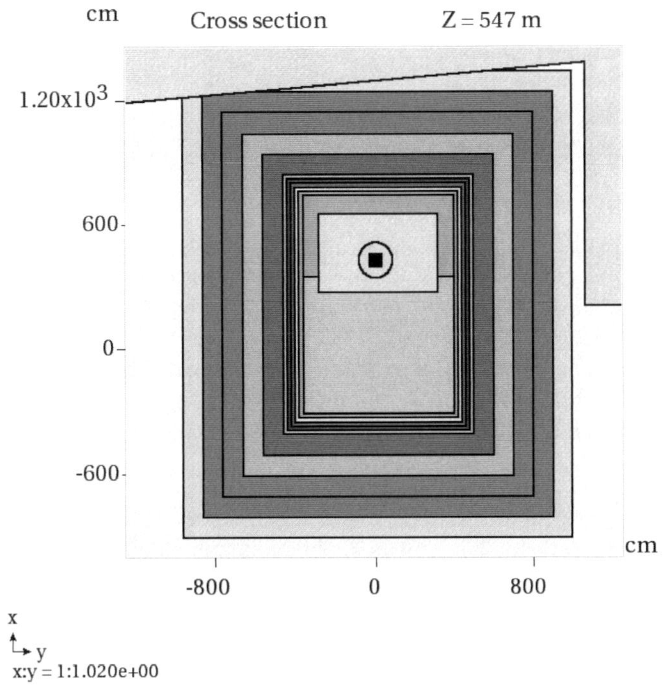

Figure 6. A cross section of the ESS accelerator and adjacent structures in the MARS15 calculation model in region 3 at z=547m with transverse bins for scoring produced radionuclides (the pink region underneath the blue region is the same concrete wall which, for technical reasons, serves as a placeholder for the beam dump in our model, see also Fig. 3).

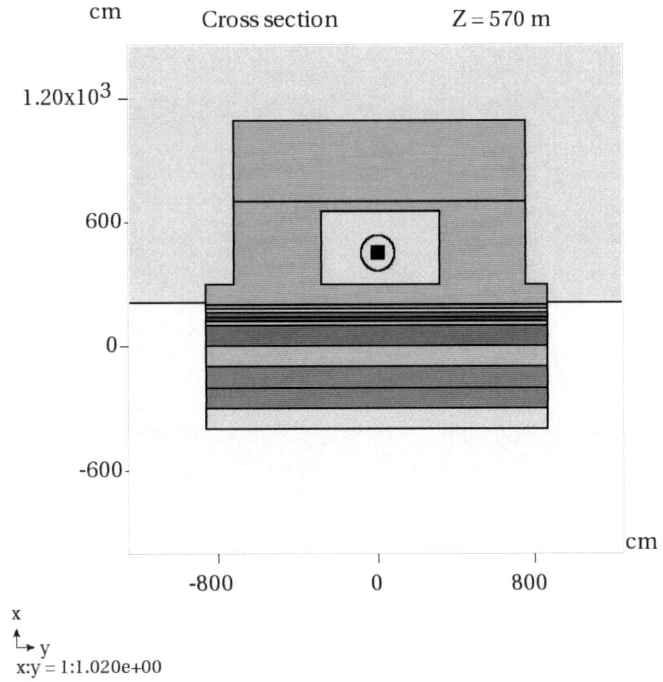

Figure 7. A cross section of the ESS accelerator and adjacent structures in the MARS15 calculation model in region 3 at z=570m with transverse bins for scoring produced radionuclides (the dark blue regions represent concrete).

IOP Conf. Series: Journal of Physics: Conf. Series **1046** (2018) 012020 doi:10.1088/1742-6596/1046/1/012020

4. Assumptions and input data

4.1. Operational scenario

The following operational scenario is assumed: in every single year, a 6000-hour beam-on period is followed by a 2760-hour beam-off period. For the 1st and 40th year of operation, the nuclide production is calculated immediately after the most recent beam-on period (in other words – at shut-down).

4.2. Beam loss

At present, for high-energy proton accelerators it is a common practice to use the so-called "1 W/m" rule derived from the hands-on maintenance conditions and suggested in Ref. [8]. It provides an average operational beam loss along the entire linac beam line except for the very first low-energy sections. The 1 W/m average homogeneous beam loss rate has been adopted at ESS [9] and is considered as a design criterion and an upper limit during normal operations.

The beam energy dependence on location along the ESS accelerator and the modelled beam loss rate at 1 W/m are shown in Fig. 8.

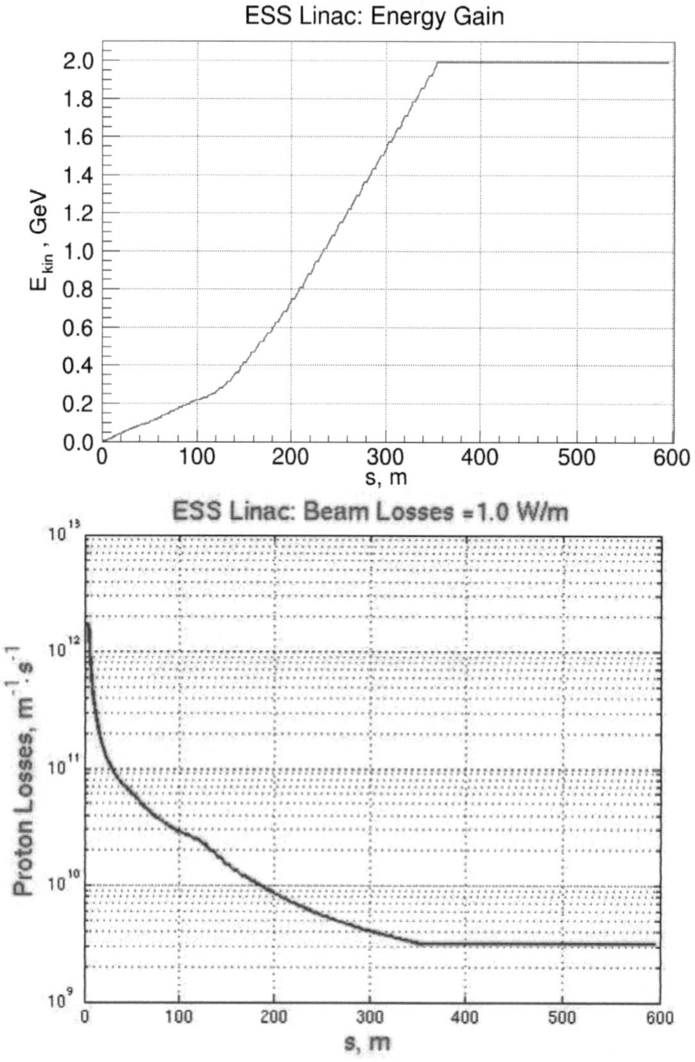

Figure 8. Primary proton beam energy as a function of location along the ESS linac (top) and the modelled normal operational beam loss rate at 1 W/m [6, 7] (bottom).

IOP Conf. Series: Journal of Physics: Conf. Series **1046** (2018) 012020 doi:10.1088/1742-6596/1046/1/012020

4.3. Shielding materials

Ordinary concrete and soil are the two main shielding materials of ESS accelerator. Chemical composition and density of both used at ESS are given in Table 1. The composition of the soil used in calculations is assumed to be a 50%-50% mixture of the deep and shallow soil sample compositions. All other materials in the MARS15 model have their standard compositions and densities. These include air, stainless steel, niobium, liquid helium, titanium, aluminium, magnet yoke steel, cables, copper, water and stub filler (90% sand mixed with 10% of poly-beads).

Table 1. ESS accelerator shielding material properties. Density of both deep and shallow ESS soil is equal to 2.0 g/cm^3, and density of ESS concrete is equal to 2.35 g/cm^3.

Chemical component	ESS Shallow Soil Sample Weight fraction (%)	ESS Deep Soil Sample Weight fraction (%)	ESS Concrete Weight fraction (%)
Water	16.4901	16.0827	3.00
Mg	0.5156	0.8278	0.19
Al	4.4515	3.6631	1.92
Si	15.45	15.45	12.51
S	0.00	0.00	0.81
K	2.0085	2.0389	1.58
Ca	0.8828	6.1268	39.27
Ti	0.2149	0.1740	0.38
Mn	0.0317	0.0283	0.21
Fe	2.2578	1.9147	4.44
Rb	0.0107	0.0095	0.00
Ba	0.0327	0.0274	0.00
Ni	0.0017	0.0015	0.12
Zr	0.0068	0.0055	0.00
O	57.00	53.00	35.57
H	1.8452	1.7997	0.00
Ce	0.0052	0.0044	0.00
Co	0.0009	0.0008	0.00
Cr	0.0040	0.0033	0.00
Cu	0.0004	0.0005	0.00
Dy	0.0005	0.0003	0.00
Eu	0.0001	0.0001	0.00
La	0.0025	0.0020	0.00
Na	0.5630	0.5706	0.00
Nd	0.0022	0.0023	0.00
P	0.0482	0.0399	0.00
Pb	0.0015	0.0008	0.00
Sc	0.0007	0.0095	0.00
Sr	0.0054	0.0158	0.00
V	0.0050	0.0041	0.00
Y	0.0016	0.0013	0.00
Yb	0.0002	0.0001	0.00
Zn	0.0039	0.0029	0.00

IOP Conf. Series: Journal of Physics: Conf. Series **1046** (2018) 012020 doi:10.1088/1742-6596/1046/1/012020

5. Results of calculations

The list of eleven most important radionuclides for groundwater activation was compiled using studies from Ref. [10]. This list comprises the following nuclides: ^{3}H, ^{7}Be, ^{22}Na, ^{24}Na, ^{32}P, ^{35}S, ^{45}Ca, ^{46}Sc, ^{54}Mn, ^{55}Fe, ^{65}Zn. In this paper, the calculated activation is provided for the radionuclides from this list. At the same time, output files of these calculations contain information on all produced radionuclides. In these calculations, statistical RMS uncertainty, 1σ, of the production rate for ^{3}H, ^{7}Be, ^{22}Na and ^{24}Na does not exceed 1%. ^{65}Zn represents produced nuclides with the lowest activity, and corresponding 1σ is approximately 8%. For the remaining produced nuclides on the list, the values of 1σ are in between 1 and 8%.

Table 2 provides the total longitudinally-integrated activity (Bq) of the relevant accounted eleven nuclides for the first transverse 1-m thick soil layer and the one for the entire soil, the 6-m thick layer. Results are given for 1 year and 40 years of operation. Table 3 provides results for the specific activity (Bq/kg) for the same scoring scheme. Table 4 provides the total activity (Bq) in the 6-m thick soil layer for the first, second and third longitudinal regions as well as the sum of the three regions after one year of operation.

The total activity over all the lateral bins in the soil, as well as total activity of the tunnel concrete and dump placeholder concrete are shown in Figure 9 and Table 5. The top ten radionuclides in the total soil volume after 40 years of operation at shut down and after 30 years of cooling are given in Table 6. The spatial distribution in the soil of the specific activity follows that of the total hadron flux above 30 MeV. Figs 10 through 12 shows lateral distributions of such a flux in the three longitudinal sections of the linac. Energy spectra of neutrons and gammas, averaged over all the ten transverse bins of the soil and over the tunnel concrete in the 3rd longitudinal section, are shown in Fig. 13.

Table 2. Activity (Bq) of radionuclides produced in the layers of soil – 1-m thick and 6-m thick – around the ESS tunnel wall immediately after 1 year and 40 years of operation (no decay).

Produced nuclide	Period of operation (yr)			
	1		40	
	Layer thickness (m)		Layer thickness (m)	
	1	6	1	6
^{3}H	1.157×10^{9}	1.285×10^{9}	1.892×10^{10}	2.101×10^{10}
^{7}Be	6.631×10^{9}	7.340×10^{9}	6.688×10^{9}	7.404×10^{9}
^{22}Na	9.188×10^{8}	1.022×10^{9}	3.930×10^{9}	4.371×10^{9}
^{24}Na	4.635×10^{10}	5.008×10^{10}	4.635×10^{10}	5.008×10^{10}
^{32}P	1.469×10^{9}	1.608×10^{9}	1.470×10^{9}	1.609×10^{9}
^{35}S	4.482×10^{8}	4.976×10^{8}	4.746×10^{8}	5.268×10^{8}
^{45}Ca	3.053×10^{7}	3.396×10^{7}	3.870×10^{7}	4.305×10^{7}
^{46}Sc	8.595×10^{8}	9.276×10^{8}	9.036×10^{8}	9.752×10^{8}
^{54}Mn	8.088×10^{8}	8.998×10^{8}	1.457×10^{9}	1.620×10^{9}
^{55}Fe	5.883×10^{8}	6.531×10^{8}	2.630×10^{9}	2.919×10^{9}
^{65}Zn	1.093×10^{6}	1.209×10^{6}	1.693×10^{6}	1.873×10^{6}

IOP Conf. Series: Journal of Physics: Conf. Series **1046** (2018) 012020 doi:10.1088/1742-6596/1046/1/012020

Table 3. Specific activity (Bq/kg) of radionuclides produced in the layers of soil – 1-m thick and 6-m thick – around the ESS tunnel wall immediately after 1 year and 40 years of operation (no decay).

Produced nuclide	Period of operation (yr)			
	1		40	
	Layer thickness (m)		Layer thickness (m)	
	1	6	1	6
^{3}H	34.4	3.95	563	64.6
^{7}Be	197	22.6	199	22.8
^{22}Na	27.3	3.14	117	13.4
^{24}Na	1380	154	1380	154
^{32}P	43.7	4.95	43.8	4.95
^{35}S	13.3	1.53	14.1	1.62
^{45}Ca	0.91	0.104	1.15	0.132
^{46}Sc	25.6	2.85	26.9	3.0
^{54}Mn	24.1	2.77	43.4	4.98
^{55}Fe	17.5	2.01	78.3	8.98
^{65}Zn	0.033	0.0037	0.05	0.0058

Table 4. Longitudinal distribution of total activity (Bq) of radionuclides produced in the 6-m thick soil layer around the ESS tunnel wall immediately after 1 year of operation (no decay).

Produced nuclide	0<Z<176m	176m<Z<490.56m	490.56m<Z<580.5m	Total
^{3}H	5.147×10^{7}	1.097×10^{9}	1.367×10^{8}	1.285×10^{9}
^{7}Be	2.065×10^{8}	6.340×10^{9}	7.936×10^{8}	7.340×10^{9}
^{22}Na	3.957×10^{7}	8.740×10^{8}	1.083×10^{8}	1.022×10^{9}
^{24}Na	4.395×10^{9}	4.105×10^{10}	4.633×10^{9}	5.008×10^{10}
^{32}P	1.041×10^{8}	1.345×10^{9}	1.591×10^{8}	1.608×10^{9}
^{35}S	2.357×10^{7}	4.215×10^{8}	5.252×10^{7}	4.976×10^{8}
^{45}Ca	1.325×10^{6}	2.934×10^{7}	3.297×10^{6}	3.396×10^{7}
^{46}Sc	8.464×10^{7}	7.575×10^{8}	8.544×10^{7}	9.276×10^{8}
^{54}Mn	4.269×10^{7}	7.631×10^{8}	9.397×10^{7}	8.998×10^{8}
^{55}Fe	3.242×10^{7}	5.521×10^{8}	6.853×10^{7}	6.531×10^{8}
^{65}Zn	0	1.069×10^{6}	1.401×10^{5}	1.209×10^{6}

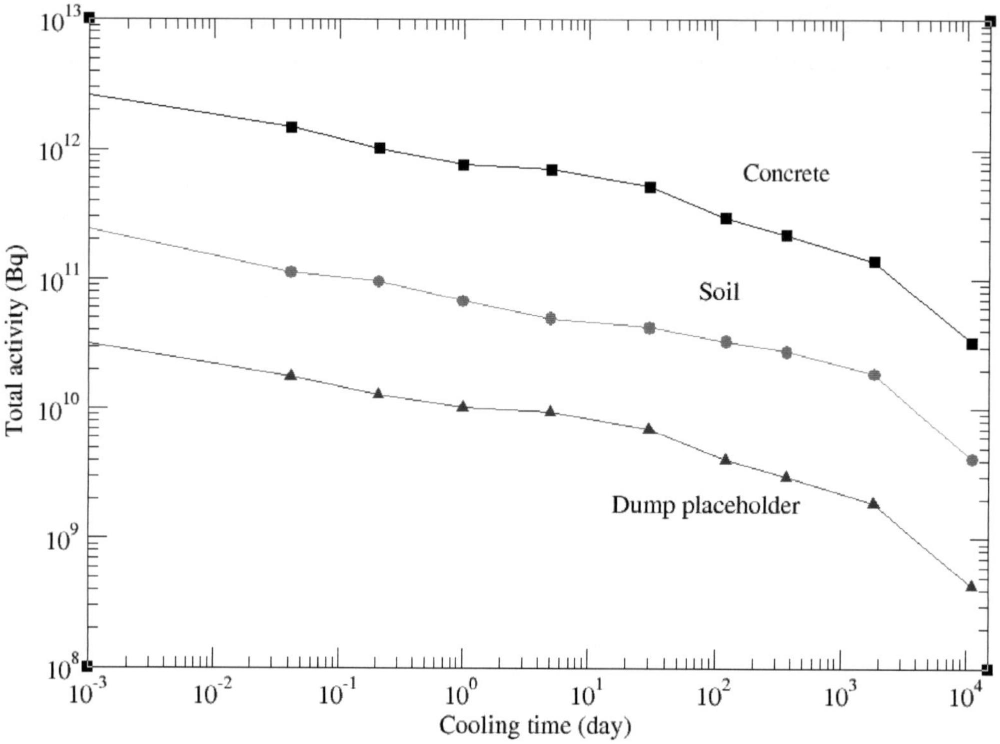

Figure 9. Total activity of all the bins in the soil, as well as total activity of the tunnel concrete and dump placeholder concrete (see Figs. 3 and 6) after 40 years of operation. Cooling time 0 corresponds to the moment of the accelerator shut down.

Table 5. Total activity (Bq) of all the bins in the soil, as well as total activity of the tunnel concrete and dump placeholder concrete after 40 years of operation.

Cooling time	Tunnel concrete	Total soil	Dump placeholder concrete
0	3.744×10^{12}	4.019×10^{11}	4.649×10^{10}
1 hour	1.484×10^{12}	1.126×10^{11}	1.751×10^{10}
5 hours	1.015×10^{12}	9.597×10^{10}	1.286×10^{10}
1 day	7.583×10^{11}	6.821×10^{10}	1.026×10^{10}
5 days	7.007×10^{11}	4.958×10^{10}	9.487×10^{9}
30 days	5.182×10^{11}	4.251×10^{10}	7.005×10^{9}
120 days	2.978×10^{11}	3.322×10^{10}	4.014×10^{9}
1 year	2.204×10^{11}	2.789×10^{10}	2.967×10^{9}
5 years	1.382×10^{11}	1.879×10^{10}	1.866×10^{9}
30 years	3.265×10^{10}	4.133×10^{9}	4.344×10^{8}

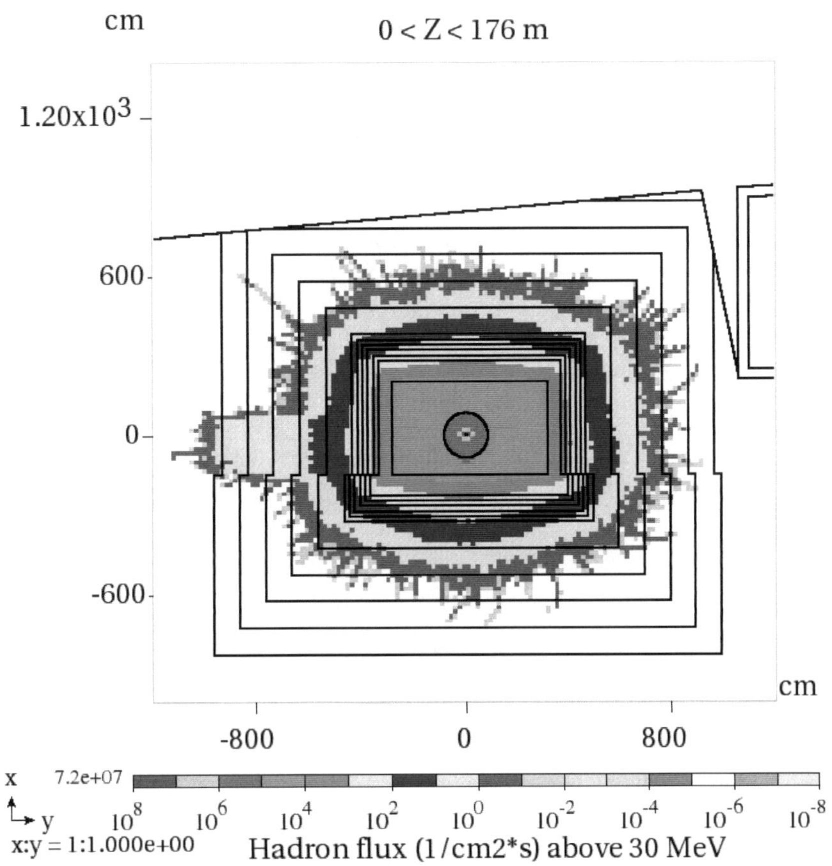

Figure 10. Transverse distribution of total hadron flux above 30 MeV averaged over the first longitudinal section of the model.

Table 6. Activity (Bq) of top ten radionuclides in all the soil bins after 40 years of operation given at shut down (left) and after subsequent 30 years of cooling (right).

Produced radionuclide	Activity	Produced radionuclide	Activity
^{15}O	1.076×10^{11}	^{3}H	3.994×10^{9}
^{28}Al	1.035×10^{11}	^{39}Ar	4.722×10^{7}
^{24}Na	5.140×10^{10}	^{14}C	4.159×10^{7}
^{3}H	2.157×10^{10}	^{41}Ca	1.907×10^{7}
^{27}Si	1.356×10^{10}	^{60}Co	1.102×10^{7}
^{13}N	1.213×10^{10}	^{44}Sc	4.059×10^{6}
^{56}Mn	1.120×10^{10}	^{44}Ti	4.059×10^{6}
^{14}O	8.259×10^{9}	^{22}Na	1.525×10^{6}
^{7}Be	7.599×10^{9}	^{55}Fe	1.511×10^{6}
^{11}C	7.168×10^{9}	^{133}Ba	1.425×10^{6}

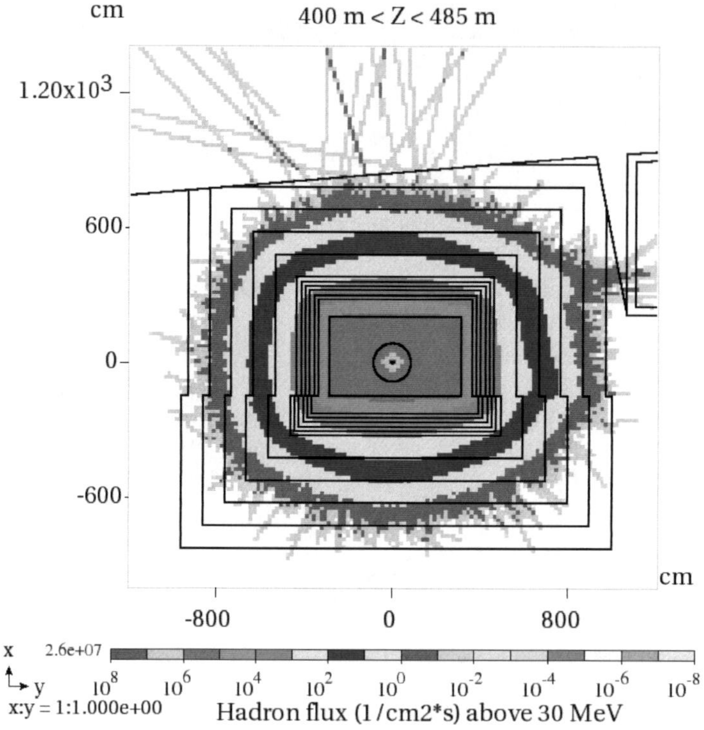

Figure 11. Transverse distribution of total hadron flux above 30 MeV averaged over a part the second longitudinal section of the model.

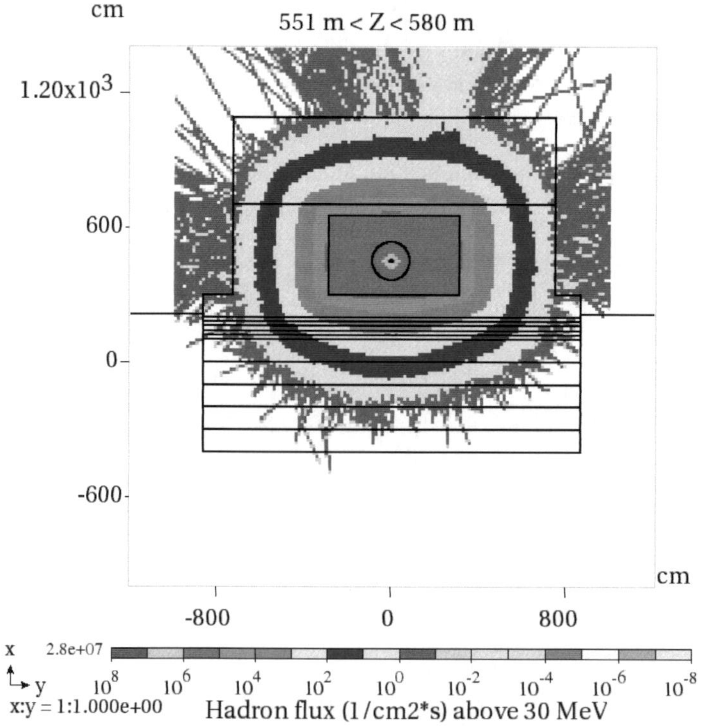

Figure 12. Transverse distribution of total hadron flux above 30 MeV averaged over a part the third longitudinal section of the model.

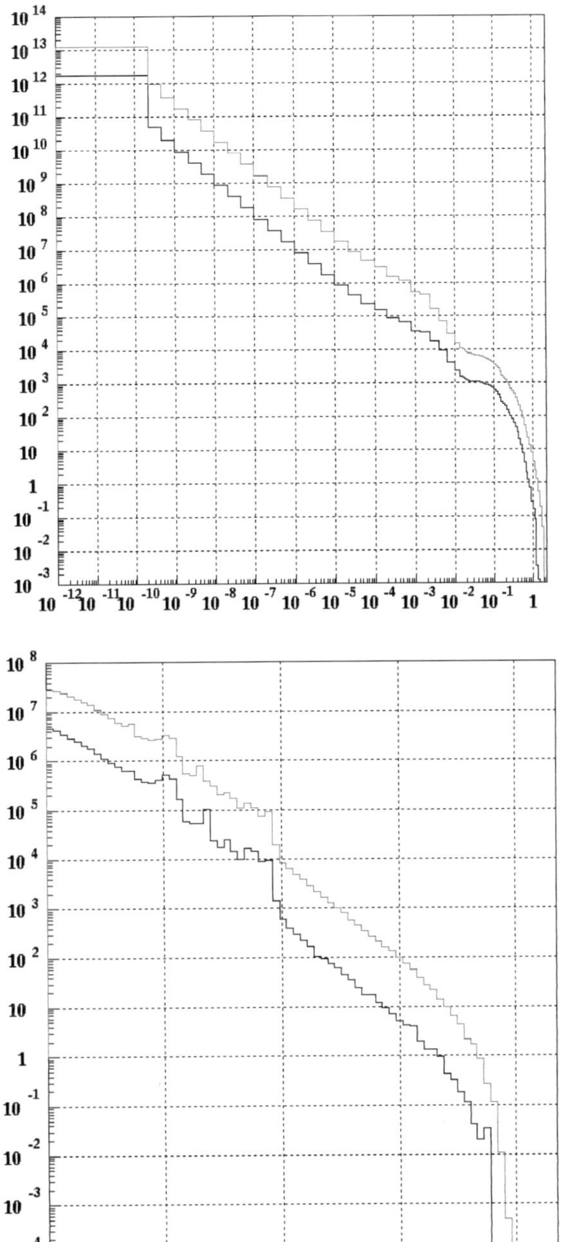

Figure 13. Energy spectra (cm^{-2} GeV^{-1} s^{-1}) vs energy (GeV) of neutrons (top) and gammas (bottom) averaged over a part of the third longitudinal section of the model (see Fig. 7) from z=550m to z=580m; the red and black lines represent the tunnel concrete and all the 10 transverse bins in soil, respectively.

6. Verification of predicted residual activation

A comparison with a relatively simple activation experiment performed at GSI, when the number of projectiles delivered to the target is well known, is presented in Ref. [3]. In this case, the MARS15 predictions agree with measured foil activation within 30% or better. This case, however, does not

IOP Conf. Series: Journal of Physics: Conf. Series **1046** (2018) 012020 doi:10.1088/1742-6596/1046/1/012020

represent all the complexity of experiments one could encounter. Therefore, comparisons with activation measured in a more complicated experiment at Fermilab are presented below.

6.1. Comparisons between measured and calculated residual activation and a normalization issue

Residual activation of various aluminum, steel and copper foils was measured in a collimation region of Main Injector at Fermilab with 8-GeV primary proton beam [11-13]. In this experiment, foil activation was measured near the beam pipe (unshielded location) and on a collimator wall which represents a steel and marble shielding approximately 60 and 15 cm, respectively, in thickness (shielded location). The beam line and foil locations are shown in Figs. 14 and 15.

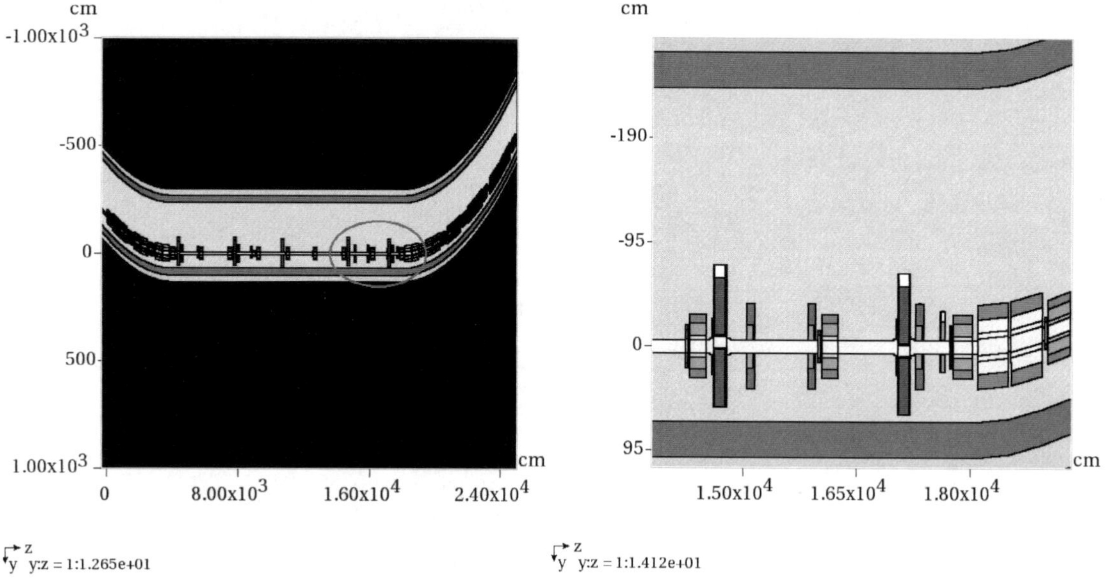

Figure 14. A plan view of the MARS15 model of the Main Injector collimation region (left) and the beam line fragment used for the measurements (right).

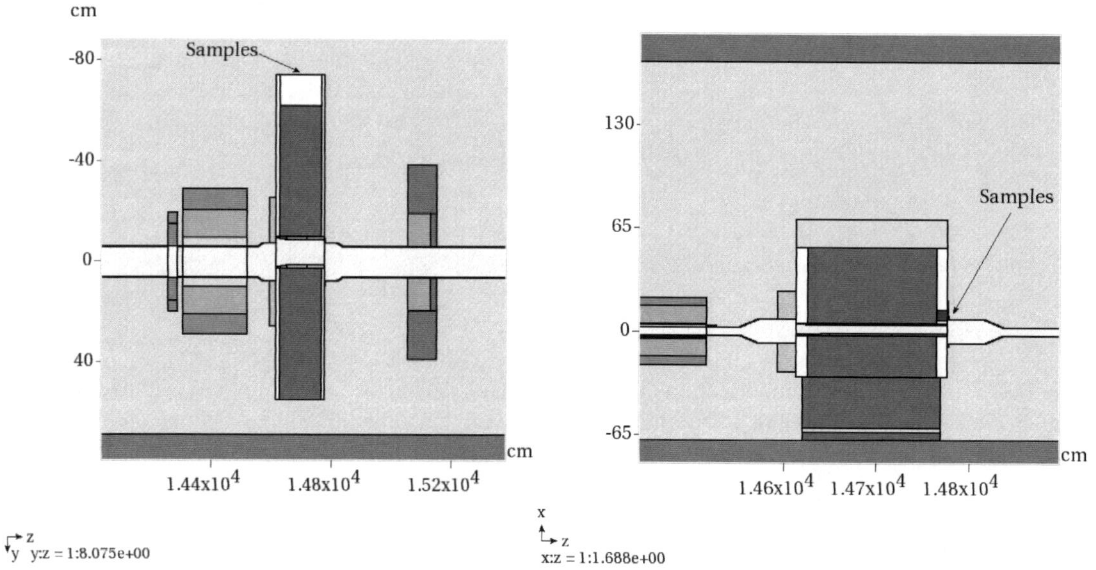

Figure 15. The beam line fragment with the activation foils: a plan view (left) and elevation view (right) that show the shielded and unshielded location, respectively.

Various irradiation and cooling times were studied. Three different methods were used in order to determine the number of protons lost on the collimator with the activation foils. The three methods provided three different numbers, with the ratios of both the highest number to the average and the average to the lowest number being approximately 1.4. So, the ratio of the highest number to the lowest one is approximately 2. In order to normalize the calculated foil activation, the average of the three estimated numbers is used. The detailed measured dependence of beam intensity vs time is used to normalize the calculated foil activation. Another source of potential uncertainty is that material composition of the steel samples is not well known, so that an average—over several available specifications— material composition is used. For aluminum and copper samples, corresponding natural mixtures are used. The comparisons between the measured and calculated foil activation are presented in Tables 7 thru 12.

6.2. Unshielded location

The steel and copper foils presented in this section were installed and removed together, so that the same irradiation and cooling down profile applies to both these foils. One can see that for most of the radionuclides in Tables 7-9 the agreement with measurement is within a factor of approximately 2, while for several radionuclides it is worse. Another conclusion from this comparison is that the calculations mostly **underestimate** the measured activation. It may be a consequence, in particular, of the above-mentioned systematic uncertainty in the number of protons that are lost on the collimator.

Table 7. Measured in Radionuclide Analysis Facility at Fermilab [11-13] and calculated specific activity (pCi/g) of various radionuclides generated in a steel foil irradiated at the unshielded location for about 93 hours.

Produced radionuclide	Experiment	Calculation	C/E
^{42}K	1510±520	1300	0.86
^{43}K	1240±170	955	0.77
^{48}Cr	863±112	1830	2.12
^{51}Cr	10700±2000	4860	0.45
^{52}Fe	904±159	645	0.71
^{52}Mn	12600±1100	11786	0.94
^{54}Mn	2200±380	808	0.37
^{24}Na	944±155	176	0.19
^{44}Sc	7680±1200	4230	0.55
^{46}Sc	463±103	218	0.47
^{47}Sc	3080±480	1635	0.53
^{48}Sc	702±129	860	1.22
^{48}V	4430±510	2776	0.63

Table 8. Measured in Radionuclide Analysis Facility at Fermilab [11-13] and calculated specific activity (pCi/g) of various radionuclides generated in an aluminium foil irradiated at the unshielded location for about 93 hours.

Produced radionuclide	Experiment	Calculation	C/E
^{7}Be	712±115	170.4	0.24
^{22}Na	355±55	115.0	0.32

IOP Conf. Series: Journal of Physics: Conf. Series **1046** (2018) 012020 doi:10.1088/1742-6596/1046/1/012020

Table 9. Measured in Radionuclide Analysis Facility at Fermilab [11-13] and calculated specific activity (pCi/g) of various radionuclides generated in a copper foil irradiated at the unshielded location for about 93 hours.

Produced radionuclide	Experiment	Calculation	C/E
^{55}Co	1510±290	2014	1.33
^{56}Co	642±145	252	0.39
^{57}Co	537±197	282	0.53
^{58}Co	3310±540	1532	0.46
^{61}Cu	47800±5000	25690	0.54
^{43}K	800±204	152	0.19
^{52}Mn	2490±240	1614	0.65
^{56}Mn	5400±720	2129	0.39
^{57}Ni	1270±250	1113	0.88
^{44}Sc	1470±280	1126	0.77
^{47}Sc	501±201	404	0.81
^{48}Sc	413±82	377	0.91
^{48}V	519±107	506	0.97

6.3. Shielded location

The steel, aluminium and copper foils presented in this section were installed and removed together, so that the same irradiation and cooling down profile applies to both these foils. One can see that for most of the radionuclides in Tables 10-12 the agreement with measurement is within a factor of approximately from 2 to 3, while for ^{56}Co it is about 4. Another conclusion from this comparison is that the calculations mostly **overestimate** the measured activation. It may be a consequence, in particular, of the above-mentioned uncertainty in the number of protons that are lost on the collimator.

Table 10. Measured in Radionuclide Analysis Facility at Fermilab [11-13] and calculated specific activity (pCi/g) of various radionuclides generated in a steel foil irradiated at the shielded location for about 45 days.

Produced radionuclide	Experiment	Calculation	C/E
^{56}Co	22.5±4.3	99.6	4.43
^{51}Cr	1870±290	3504	1.87
^{52}Mn	439±40	1548	3.53
^{54}Mn	639±97	1520	2.38
^{46}Sc	58.3±7.7	145.6	2.50
^{47}Sc	44.9±11.6	100.1	2.22
^{48}V	358±39	880	2.46

Table 11. Measured in Radionuclide Analysis Facility at Fermilab [11-13] and calculated specific activity (pCi/g) of various radionuclides generated in an aluminium foil irradiated at the shielded location for about 45 days.

Produced radionuclide	Experiment	Calculation	C/E
^{22}Na	45.8±11.8	41.7	0.91
^{24}Na	1270±210	1102	0.87

Table 12. Measured in Radionuclide Analysis Facility at Fermilab [11-13] and calculated specific activity (pCi/g) of various radionuclides generated in a copper foil irradiated at the shielded location for about 45 days.

Produced radionuclide	Experiment	Calculation	C/E
^{56}Co	78.7±8.2	307	3.90
^{57}Co	138±18	324	2.35
^{58}Co	707±107	1844	2.61
^{60}Co	22.6±3.5	50.4	2.23
^{51}Cr	131±51	256	1.95
^{59}Fe	70.0±9.4	50.1	0.72
^{52}Mn	43.9±5.2	128	2.92
^{54}Mn	49.3±8.4	85.4	1.73
^{48}V	39.1±5.0	36.6	0.94

7. Conclusions

Spatial distributions of radionuclides produced in the soil outside the ESS accelerator tunnel wall were calculated for a realistic operational scenario. The distributions are for subsequent analysis of radionuclide migration by a dedicated expert group. A comparison between predicted and measured foil activation, performed for a similar radiation environment, for most of the studied radionuclides revealed an agreement between calculated and measured residual foil activation within a factor of from one to three.

Acknowledgements

This document was prepared using the resources of the Fermi National Accelerator Laboratory (Fermilab), a U.S. Department of Energy, Office of Science, HEP User Facility. Fermilab is managed by Fermi Research Alliance, LLC (FRA) acting under contract No. DE-AC02-07CH11359.

This research used a computing award allocation from the ASCR Leadership Computing Challenge at the Argonne Leadership Computing Facility, which is a DOE Office of Science User Facility supported under Contract DE-AC02-06CH11357.

This work was supported by a CRADA agreement NO FRA-2016-0058 between Fermi Research Alliance, LLC and European Spallation Source ERIC.

Authors are grateful to B. C. Brown of Fermilab for helpful discussions.

References

[1] https://europeanspallationsource.se/accelerator
[2] Mokhov N V 1995 The MARS Code System User's Guide *Fermilab-FN-628*
[3] Mokhov N V *et al* 2014 *Prog. Nucl. Sci. Technol.* **4** 496
[4] https://mars.fnal.gov
[5] Aarnio P 1998 Decay and Transmutation of Nuclides *CERN CMS-NOTE-1998/086*

[6] Mokhov N, Rakhno I, Tropin I, Eidelman Yu and Tchelidze L 2016 *ESS accelerator prompt radiation shielding assessment ESS-0052477, rev. 2*
[7] Mokhov N V, Eidelman Yu I, Rakhno I L, Tchelidze L, Tropin I S 2016 MARS15 simulation of radiation environment at the ESS linac *Fermilab-Conf-16-543-APC*
[8] Krivosheev O E and Mokhov N V 2000 Tolerable Beam Loss at High-Intensity Proton Machines *Fermilab-Conf-00/192*
[9] ESS hands on maintenance conditions for ESS accelerator *ESS-0008351, rev. 2*
[10] Thomas R H and Stevenson G R 1988 Radiological Safety Aspects of the Operation of Proton Accelerators *IAEA TERS 283*
[11] Brown B C 2011 Analysis procedures for Al activation studies *Fermilab Beams-doc 3980*
[12] Brown B C 2015 Calibration of BLM response to proton loss *Fermilab Beams-doc 4867*
[13] Brown B C 2017 Activation of steel and copper samples in the Main Injector collimation region *Fermilab Beams-doc 4046 ver. 2.27*